TO MY WIFE
E. J. SCOVELL

Published by arrangement with Kluwer Academic Publishers B.V.
The University of Chicago Press, Chicago 60637
First published in 1958

University of Chicago Press Edition 2000
Printed in the United States of America
09 08 07 06 05 04 03 02 01 00   1 2 3 4 5
ISBN: 0-226-20638-6 (paperback)

Library of Congress Cataloging-in-Publication Data

Elton, Charles S. (Charles Sutherland), 1900–
The ecology of invasions by animals and plants / by Charles S. Elton ; with a foreword by Daniel Simberloff.—University of Chicago Press ed.
p. cm.
Originally published: London : Methuen, c1958. With a new foreword.
Includes bibliographical references (p.).
ISBN 0-226-20638-6
1. Ecology. 2. Biogeography. 3. Animal introduction. 4. Plant introduction.
I. Title.

QH541.E4 2000   99-052582
577'.18—dc21

∞ The paper used in this publication meets the Minimum requirements of the American National Standard for Information Sciences—Permanence of Paper for Printed Library Materials, ANSI Z39.48-1992.

Contents

Foreword

Charles Elton was a founder of ecology.[1,2] His *Animal Ecology* (1927), reprinted nine times through 1966, is a classic of population and community ecology and inspired much current research in these fields. A leader in the study of subjects that range from species (such as lemming and snowshoe hare cycles) through ecosystems (such as the nature of food chains and food webs), Elton was particularly interested in how species interact with one another. Why were there 2,500 animal species in a small woodland he studied, how could they all coexist, and why weren't there 5,000? This interest, plus his keen naturalist's eye, led him to an early concern with invasions by introduced species. His first publication on this subject was a note in the *Times* of London in 1933.[3] However, the book that founded a whole field of research, *The Ecology of Invasions by Animals and Plants,* arose in an unorthodox fashion. In 1957, Elton gave three BBC radio broadcasts, entitled "Balance and Barrier." Within a year, he had expanded these ideas into what was to become a bible for practitioners of a burgeoning new science: invasion biology. Virtually every monograph in the field cites *The Ecology of Invasions* as inspirational and still timely. The most influential current effort, Mark Williamson's *Biological Invasions,* says that, "Charles Elton, the father of animal ecology in Britain, with his broadcasts and book (1958) drove much of the interest and understanding of invasions in our lifetime."[4] Yet, the book is aimed at a lay audience, replete with apt references to H. G. Wells, Buffalo Bill, Walt Whitman, Robert Browning, Charlie Chaplin, and Peter Brueghel the Elder.

The science of invasion biology has three basic components, and, though previous authors had recognized some of them, Elton was the first to unite them and recognize their interdependence. Further, much as Darwin did in *The Origin of Species,* Elton used a plethora of dramatic examples and vivid metaphors to draw patterns and to establish a cohesive framework to encompass many disparate phenomena. These are the components. First, over hundreds of millions of years, the plant and animal communities of the different continents have come to be very distinct from one another. Second, human trade and travel are rapidly obliterating these distinctions. Third, this process has grave implications for the conservation of diversity.

Elton's examples remain timely. In the United States alone, European gypsy moths continue to ravage northeastern forests and are spreading southward while Asian gypsy moths periodically make small beachheads in the Pacific Northwest. The South American nutria threatens the very existence of coastal wetlands in Louisiana and Maryland. Asian chestnut blight, which removed almost all adult chestnut trees from 225 million acres of the eastern United States, continues to thwart all efforts to reestablish what had been a dominant plant. Argentine ants eliminate native insects from increasingly wide areas of California and Hawaii. Chinese mitten crabs have become a costly problem in California, clogging fish screens and devouring native fish fry and invertebrates. The central European balsam woolly adelgid has almost eradicated Fraser fir, the dominant high elevation tree, from most of the southern Appalachians, leaving thousands of dead, rotting trees covering the highest peaks. North Atlantic sea lampreys, having driven three endemic fishes of the Great Lakes to extinction, still kill many lake trout and other fishes and impede the rebuilding of established populations and fisheries, despite an annual outlay of $10,000,000 to combat them. The hybrid cordgrass that threatened English mudflats is still a major weed there, and its eastern North American parental species has now invaded California and Washington, where it is a huge problem and the target of extensive control operations. All these species are big newspaper stories today and also topics of intense research and discussion among scientists and managers. Elton's examples from elsewhere are similarly apt today. For instance, the German wasp has spread still farther in New Zealand, though in some forests it has largely been replaced by a later, related European arrival, the common wasp. These wasps attack native insects and are even suspected of competing for nectar with a threatened endemic parrot.

What is even more remarkable about Elton's book is that he predicted not only that such disastrous invasions would increase in number, but the nature of some of the species that have recently emerged as horrors. European zebra mussels first appeared in North America in the mid-1980s and have become a multibillion dollar economic problem and a threat to numerous native molluscs. They arrived first in the Great Lakes on fouled hulls or in ballast water, then spread as the adults fouled boats and barges. Elton highlighted fouling and ballast water as particularly likely to spread aquatic and marine organisms. Additionally, he emphasized that species that make their livings very differently from all the natives are most likely to wreak havoc upon introduction. The zebra mussel fills an "empty niche" in North America; its combination of byssal threads (to attach to rocks, plants, and other stable surfaces), active water filtration, and larvae that feed in the plankton is found in no native species and probably accounts for its dramatic impact.

Elton discussed some successes and failures of biological control—the deliberate introduction of what he called "counterpests" to battle existing invaders. And he also discussed the invasion of the giant African snail, *Achatina fulica,* on Hawaii and other Pacific islands, as well as biological control projects just underway then against this invader, using imported predatory snails. Separately, he pointed out that other biological control agents, wasps introduced to control exotic agricultural pests, had probably eliminated native, nontarget moth species in Hawaiian forests. And throughout his book, Elton depicted the unintended consequences of species that humans had deliberately introduced. He was prophetic. One predatory snail introduced to Hawaii and elsewhere as he was writing—the rosy wolf snail (*Euglandina rosea*), a native of Florida and Central America—ended up not only failing to control the giant African snail but causing the extinction of at least 30 species and subspecies of endemic snails on Pacific islands, a hecatomb rarely if ever matched by a single deliberate human action.

Until the advent of the molecular revolution, biologists often debated the proposition that all academic study of evolution was simply a footnote to Darwin. Could one make a similar claim for Elton and invasion biology? Elton himself lamented the fact that his book was largely a compendium of examples. Today there is always a new catastrophic-pest-of-the-week causing ecological, economic, or aesthetic turmoil. The Asian long-horned beetle in New York and Chicago, the Asian tiger mosquito and tropical soda apple in the southeastern United States, the tropical "killer alga" (*Caulerpa taxifolia*) in the Mediterranean, African "killer bees" in South America and now North America, South American miconia trees in Hawaii, the northern Pacific seastar (*Asterias amurensis*) in Australia, a western Atlantic comb jellyfish (*Mnemiopsis leidyi*) in the Black Sea and Azov Sea—all are post-Elton invaders. Many other species were present when he wrote but have since proliferated and spread to become much greater components of the biota—Australian paperbark (melaleuca) and Brazilian pepper trees in Florida, Asian kudzu and cogon grass in the Southeast, the Formosan termite in the Southeast and California, the brown tree snake from the Admiralty Islands in Guam, the North American gray squirrel in Great Britain, European purple loosestrife in much of the United States, the Nile perch and South American water hyacinth in Lake Victoria. What generalizations can be drawn from such a catalog?

Elton clearly foresaw some patterns that have since been more comprehensively confirmed and are now the reigning tenets of invasion biology (all are well treated in Williamson's *Biological Invasions*). First, most introduced species don't survive, and most that do persist do not invade native communities. Elton perceived "ecological resistance"—native competitors, predators, parasites, and diseases of a newly arrived species—as the explanation for both statistics, although he pointed to an occasional role for other factors, such as habitats and/or climates. Nowadays, most invasion biologists would probably agree with him, though some would claim a large role for chance in determining the immediate survival (but not the spread) of a few individuals of a new invader. Second, islands—even large ones like the North and South Islands of New Zealand—have been particularly devastated by introduced species. Elton attributed this pattern to the generally reduced number of native species on islands and the absence of natives in certain entire "ecological niches" (like ground-dwelling mammalian predators). Some current practitioners would contest the first point (see below); all would agree with the second. Third, the spatial spread of invaders is often remarkably regular. Elton noted this feature but did not explore the underlying mathematics; many mathematicians and ecologists have subsequently done so. Fourth, Elton observed that introduced invaders have a bewildering variety of impacts (including species extinctions) and, in sum, are causing ecological changes as dramatic as those of the geological past, such as mass extinctions. We now believe that introduced species are the second greatest cause of species extinction and endangerment (following habitat destruction), both in the United States and worldwide, and ecologists have documented extensive disruptions at the population, species, community, and ecosystem levels.

Finally, Elton lamented the low level of predictability—which invaders will survive and become huge pests? And he hoped invasion biology would advance rapidly beyond the example and pattern stage. Most current practitioners would probably claim some progress has been made on both counts, but it is widely conceded that predictability is still low and that the field is still dominated by case studies and attempts to draw generalizations. Controlled, replicated experiments are increasing, but slowly.

What did Elton get wrong, and what did he leave for us to do other than mop-up operations? Only one theme in *The Ecology of Invasions* draws outright scientific criticism. Although he noted that invaders sometimes get into natural habitats, Elton felt that sites disturbed by humans—roadsides, agricultural fields, and the like—received much more than their share. Most researchers today would agree with this contention but many would question his explanation. Elton saw this feature of disturbed habitats, as well as the greater damage on islands than on a mainland, as a consequence of an inherently greater susceptibility to invasion by communities with small numbers of species. This idea, which he conceded was an untested theory, dovetailed with his emphasis on ecological resistance as the key to failed or minor invasions. This notion that invasibility depends on the number of native species is a version of a diversity-stability relationship that constituted received wisdom in academic ecology in the 1950s and early 1960s—the idea was that there is a strict relationship between increased stability and increased number of species. This notion faced hard questioning in the 1970s and was quite discredited by 1980,[5] though it lives on among much of the public and has recently recrudesced, somewhat transmogrified, in a widespread (and controversial) attempt to show that "healthier" function of ecosystems is inevitably fostered by increased numbers of species. What Elton largely ignored is the different numbers of species and individual invaders ("propagule pressure" in scientific parlance) to which different habitats and sites are subjected. Many disturbed sites and islands have received greater numbers of immigrant species, and sometimes greater numbers of individuals of many of these species, than intact mainland natural areas, and this influx may affect the number of invading species that survive and their impact.

The subsequent trajectories of Elton's examples, as noted above, showed that they were usually well chosen and occasionally remarkably farsighted. History shows that a few were incorrect. The most egregious, tellingly, is one of the few introduced species that Elton cites as beneficial, the Asian multiflora rose, widely planted in the eastern United States by the Soil Conservation Service and other federal and state agencies as a form of living fence, as food and shelter for wildlife, and for erosion control. By the early 1960s this plant was widely recognized as an invasive pest, and it has since been the target of removal campaigns involving tractors, bulldozers, herbicides, goats, and imported insects.[6]

Also, a few themes Elton either downplayed or omitted are now critical components of the current research agenda in invasion biology. Although he pointed to the invasion of England by a hybrid cordgrass, of which one parent was native and the other a noninvasive exotic, Elton in 1958 did not recognize the massive genetic changes many native species are undergoing by virtue of hybridization with introduced relatives. Some native species have been hybridized to the point of genetic extinction, such as the Texas fish *Gambusia amistadensis,* done in by massive interbreeding with introduced mosquito fish, and the New Zealand grey duck, massively interbred with exotic North American mallards. Furthermore, except for introduced insects' evolving resistance to pesticides, Elton did not concern himself with evolution of introduced species in their new homes. He was an ecologist, not an evolutionist.

He was also a zoologist, and, though he dealt with plant invasions, they were distinctly secondary to animals. In fact, at least as many plants as animals become disastrous invaders, because some plants manage to overgrow an entire area, choking out virtually every other plant and creating monocultures unsuitable as habitat for native animals. Elton cited a few synergisms between animal species (especially aphids and ants) that make the combined invasion worse than either would have been alone. Many introduced animals form synergistic relationships with introduced plants. For example, the zebra mussel, clarifying water by its filtering activities, favors invasion by large-leaved aquatic plants, such as Eurasian water milfoil. The latter species, in turn, provide new substrate for settlement of mussel larvae. Such interactions among invaders are now becoming a focus of intense research activity. So is the phenomenon of a lag time in invasion impact. Some species, such as Brazilian pepper in Florida, remain extant but relatively innocuous for many years, then rapidly spread to become pests of major proportion. Elton did not recognize this problem.

So, though Elton sketched the broad outlines of the scientific questions asked and the nature of the research to answer them, there **is** still much work for invasion biologists to do. Just as he predicted, the practical consequences of this problem have grown enormously. (A recent estimate of damage in the U.S. alone is over $120 billion annually!) Though he enumerated the ways to deal with the problem—keeping exotic species out, eradicating them before they spread once they get in, managing them at acceptable levels if they can't be eradicated—much research is needed to make these tools effective. Anyone wishing to understand the problem, from ordinary citizens through specialized researchers, can profit from reading (or rereading) *The Ecology of Invasions*. It imparts a wealth of timely information, while Elton's erudition and dry wit are periodically dazzling. A writer who can describe oysters as "a kind of sessile sheep" and characterize advances in quarantine methods by the proposition that "no one is likely to get into New Zealand again accompanied by a live red deer" is more than just a scientist pointing out an important unrecognized problem.

DANIEL SIMBERLOFF

NOTES

1. SOUTHWOOD, R. and J. R. CLARKE, 1999. Charles Sutherland Elton (obituary). *Biographical Memoirs of Fellows of the Royal Society,* in press.
2. CROWCRAFT, P. 1991. *Elton's Ecologists.* Chicago: University of Chicago Press.
3. Alien invaders. *The Times* (London), May 6, 1933, pp. 13—14.
4. WILLIAMSON, M. 1996. *Biological Invasions,* p. 2. London: Chapman & Hall.
5. SHRADER-FRECHETTE, K. S. and E. D. MC COY, 1993. *Method in Ecology.* Cambridge: Cambridge University Press.
6. ARMINE, J. W., JR., and T. A. STASNY, 1993. Biocontrol of multiflora rose. In *Biological Pollution,* ed. B. N. McKnight 9–21. Indianapolis: Indiana Academy of Science.

Illustrations

PLATES

FIGURES IN THE TEXT

Preface

In this book I have tried to bring together ideas from three different streams of thought with which I have been closely concerned during the last thirty years or so. The first is faunal history, usually regarded as a purely academic subject, but to some of whose events can be traced a number of the serious dislocations taking place in the world today. The second is ecology, particularly the structure and dynamics of populations. The third is conservation. I first published a few ideas about the significance of invasions in 1943, in a war-time review called *Polish Science and Learning*, under the title of 'The changing realms of animal life'. Since then I have had the opportunity to think pretty hard about conservation, while taking part in the planning and development of the Nature Conservancy. In March 1957 I gave three broadcasts in the B.B.C.'s Third Programme, under the title of 'Balance and Barrier'. These were subsequently printed in *The Listener* (1957, Vol. 57, pp. 514–15, 556–7, 596–7, and 600). The present book is essentially an expansion of these. I am extremely grateful to Mr James C. Thornton and Dr John Simons for advice and help in planning and giving these talks.

In preparing this book I have had invaluable assistance from the staff of the Bureau of Animal Population. Miss C. M. Gibbs typed the fair copy. Miss M. Nicholls has given me much advice on bibliographical matters. And Mr Denys Kempson has employed his superlative skill at photography in copying and printing the 101 illustrations. Without his help particularly I could not have made the book in its present form.

For permission to reproduce illustrations I am very grateful to a number of people and institutions, who are individually acknowledged in the legends under them. I want to thank Mrs M. J. Thornton and Mr J. S. Watson very much for allowing me the use of original photographs. The following have given invaluable help in getting me the use of other unpublished photographs: Dr Paul DeBach, Citrus Experiment

Station, University of California; Mr F. H. Jacob, Plant Pathology Laboratory, Ministry of Agriculture, Fisheries and Food; Dr R. F. Morris, Forest Biology Laboratory, Science Service, Canadian Department of Agriculture; Miss P. Sichel, National Maritime Museum, Greenwich; Dr Edward Graham and Dr William Van Dersal, U.S. Soil Conservation Service.

I am obliged to Dr W. E. Swinton, British Museum (Natural History), for some information about dinosaurs, and to Dr Erling Christophersen for information about plant species on Tristan da Cunha.

I have found useful references in a paper by Marston Bates (1956), 'Man as an agent in the spread of organisms' (in *Man's role in changing the face of the earth*, ed. by W. L. Thomas and others, Chicago, pp. 788–804). This is the only recent general review of the subject of invasions that I have seen.

The life-group pictures in Chapter 2, borrowed from Alfred Russel Wallace's great book *The Geographical Distribution of Animals*, are included not only for their own merit, but because I discovered that only two members of a large class of advanced zoology students had ever read the book. I have kept his Latin names without any attempt to bring them up to date, but have only used the genera and not the species.

I am grateful to my wife for reading the whole of this book before publication and for making most valuable suggestions.

Bureau of Animal Population,
Department of Zoological Field Studies,
Botanic Garden, Oxford.
24 *July* 1957

CHAPTER ONE

The Invaders

Nowadays we live in a very explosive world, and while we may not know where or when the next outburst will be, we might hope to find ways of stopping it or at any rate damping down its force. It is not just nuclear bombs and wars that threaten us, though these rank very high on the list at the moment: there are other sorts of explosions, and this book is about ecological explosions. An ecological explosion means the enormous increase in numbers of some kind of living organism—it may be an infectious virus like influenza, or a bacterium like bubonic plague, or a fungus like that of the potato disease, a green plant like the prickly pear, or an animal like the grey squirrel. I use the word 'explosion' deliberately, because it means the bursting out from control of forces that were previously held in restraint by other forces. Indeed the word was originally used to describe the barracking of actors by an audience whom they were no longer able to restrain by the quality of their performance.

Ecological explosions differ from some of the rest by not making such a loud noise and in taking longer to happen. That is to say, they may develop slowly and they may die down slowly; but they can be very impressive in their effects, and many people have been ruined by them, or died or forced to emigrate. At the end of the First World War, pandemic influenza broke out on the Western Front, and thence rolled right round the world, eventually, not sparing even the Eskimos of Labrador and Greenland, and it is reputed to have killed 100 million human beings. Bubonic plague is still pursuing its great modern pandemic that started at the back of China in the end of last century, was carried by ship rats to India, South Africa, and other continents, and now smoulders among hundreds of species of wild rodents there, as well as in its chief original

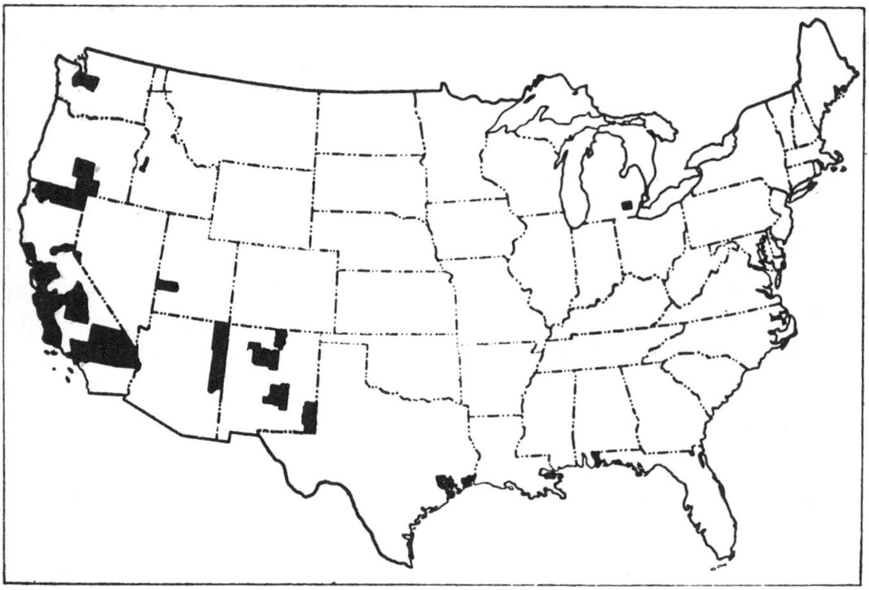

FIG. 1. Counties in the United States where plague has occurred in man. (From V. B. Link, 1955.)

home in Eastern Asia. In China it occasionally flares up on a very large scale in the pneumonic form, resembling the Black Death of medieval Europe. In 1911 about 60,000 people in Manchuria died in this way. This form of the disease, which spreads directly from one person to another without the intermediate link of a flea, has mercifully been scarce in the newly invaded continents. Wherever plague has got into natural ecological communities, it is liable to explode on a smaller or larger scale, though by a stroke of fortune for the human race, the train of contacts that starts this up is not very easily fired. In South Africa the gerbilles living on the veld carry the bacteria permanently in many of their populations. Natural epidemics flare up among them frequently. From them the bacteria can pass through a flea to the multimammate mouse; this species, unlike the gerbilles, lives in contact with man's domestic rat; the latter may become infected occasionally and from it isolated human cases of bubonic plague arise.[4] These in turn may spread into a small local

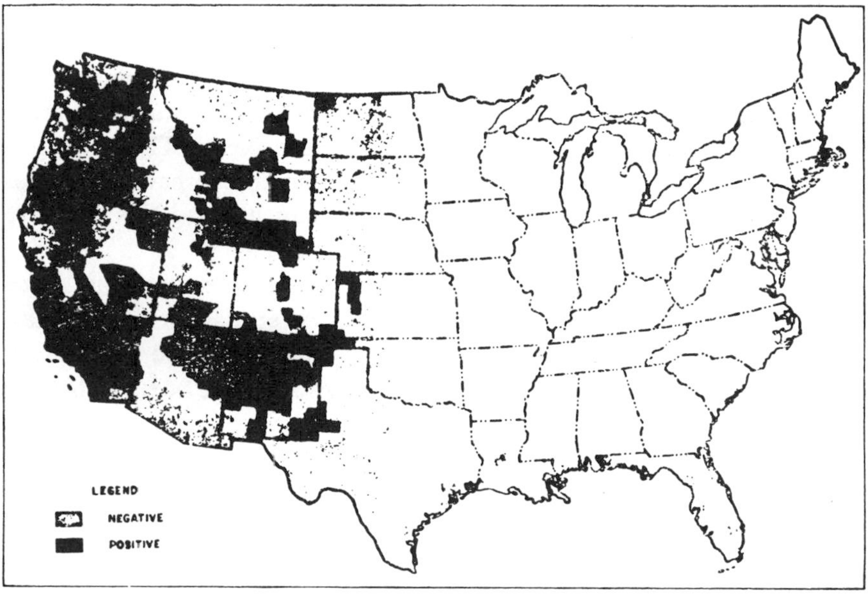

FIG. 2. Counties in the United States where plague has occurred in rodents. (From V. B. Link, 1955.)

epidemic, but often do not. In the United States and Canada a similar underworld of plague (with different species in it) is established over an immense extent of the Western regions (Pls. 1–3, Figs. 1–2), though few outbreaks have happened in man.[22] Here, then, the chain of connexions is weaker even than in South Africa, though the potentiality is present. Although plague-stricken people and plague-infected rats certainly landed from ships in California early this century, it is still possible that the plague organism was already present in North America. Professor Karl Meyer, who started the chief ecological research on sylvatic plague there, says: 'The only conclusion one can draw is that the original source and date of the creation of the endemic sylvatic plague area on the North American Continent, inclusive [of] Canada, must remain a matter of further investigation and critical analysis.'[24]

Another kind of explosion was that of the potato fungus from Europe that partly emptied Ireland through famine a hundred years ago. Most people have had experience of some kind of invasion by a foreign species, if only on a moderate scale. Though these are silent explosions in themselves, they often make quite a loud noise in the Press, and one may come across banner headlines like 'Malaria Epidemic Hits Brazil', 'Forest Damage on Cannock Chase', or 'Rabbit Disease in Kent'. This arrival of rabbit disease—myxomatosis—and its subsequent spread have made one of the biggest ecological explosions Great Britain has had this century, and its ramifying effects will be felt for many years.

But it is not just headlines or a more efficient news service that make such events commoner in our lives than they were last century. They are really happening much more commonly; indeed they are so frequent nowadays in every continent and island, and even in the oceans, that we need to understand what is causing them and try to arrive at some general viewpoint about the whole business. Why should a comfortably placed virus living in Brazilian cotton-tail rabbits suddenly wipe out a great part of the rabbit populations of Western Europe? Why do we have to worry about the Colorado potato beetle now, more than 300 years after the introduction of the potato itself? Why should the pine looper moth break out in Staffordshire and Morayshire pine plantations two years ago? It has been doing this on the Continent for over 150 years; it is not a new introduction to this country.

The examples given above point to two rather different kinds of outbreaks in populations: those that occur because a foreign species successfully invades another country, and those that happen in native or long-established populations. This book is chiefly about the first kind—the invaders. But the interaction of fresh arrivals with the native fauna and flora leads to some consideration of ecological ideas and research about the balance within and between communities as a whole. In other words, the whole matter goes far wider than any technological discussion of pest control, though many of the examples are taken from applied ecology. The real thing is that we are living in a period of the world's history when the mingling of thousands of kinds of organisms from different parts of the world is setting up terrific dislocations in nature. We are seeing huge changes in the natural population balance of the world. Of course, pest control is very important, because we have to preserve our living resources and protect ourselves from diseases and the consequences of economic dislocation. But one should try to see the whole matter on a much broader canvas than that. I like the words of Dr Johnson: 'Whatever makes the past, the distant, or the future, predominate over the present, advances us in the dignity of thinking beings.'[16] The larger ecological explosions have helped to alter the course of world history, and, as will be shown, can often be traced to a breakdown in the isolation of continents and islands built up during the early and middle parts of the Tertiary Period.

In order to focus the subject, here are seven case histories of species

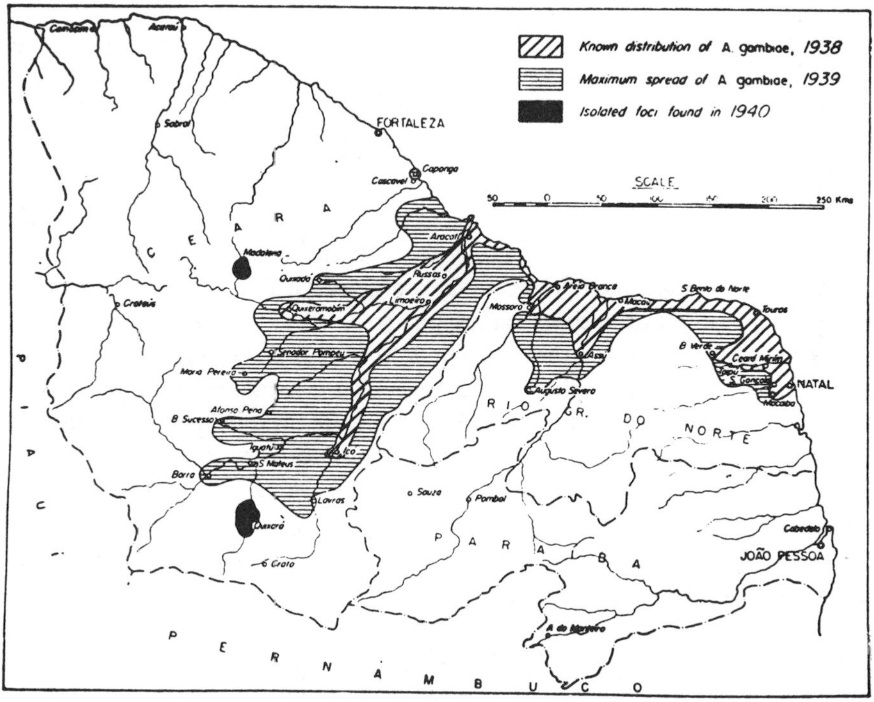

FIG. 3. Distribution areas of the African malaria mosquito, *Anopheles gambiae*, in Brazil in 1938, 1939, and 1940. Eradication measures had made it extinct in South America after this. (From F. L. Soper and D. B. Wilson, 1943.)

which were brought from one country and exploded into another. About 1929, a few African mosquitoes accidentally reached the north-east corner of Brazil, having probably been carried from Dakar on a fast French destroyer. They managed to get ashore and founded a small colony in a marsh near the coast—the Mosquito Fathers as it were. At first not much attention was paid to them, though there was a pretty sharp outbreak of malaria in the local town, during which practically every person was infected. For the next few years the insects spread rather quietly along the coastal region, until at a spot about 200 miles farther on explosive malaria blazed up and continued in 1938 and 1939, by which time the mosquitoes were found to have moved a further 200 miles inland up the Jaguaribe River valley (Fig. 3). It was one of the worst epidemics that Brazil had ever known, hundreds of thousands of people were ill, some twenty thousand are believed to have died, and the life of the countryside was partially paralysed. The biological reasons for this disaster were horribly simple: there had always been malaria-carrying mosquitoes in the country, but none that regularly flew into houses like the African species, and could also breed so successfully in open sunny pools outside the shade of the forest. Fortunately both these habits made control possible, and the Rockefeller Foundation combined with the Brazil government to wage a really astounding campaign, so thorough and drastic was it, using a staff of over three thousand people who dealt with all the breeding sites and sprayed the inside of houses. This prodigious enterprise succeeded, at a cost of over two million dollars, in completely exterminating *Anopheles gambiae* on the South American continent within three years.[28]

Here we can see three chief elements that recur in this sort of situation. First there is the historical one:—this species of mosquito was confined to tropical Africa but got carried to South America by man. Secondly, the ecological features—its method of breeding, and its choice of place to rest and to feed on man. It is quite certain that the campaign could never have succeeded without the intense ecological surveys and study that lay behind the inspection and control methods. The third thing is the disastrous consequences of the introduction. One further consequence was that quarantine inspection of aircraft was started, and in one of these they discovered a tsetse fly, *Glossina palpalis*, the African carrier of sleeping sickness in man, and at the present day not found outside Africa.[28]

The second example is a plant disease. At the beginning of this century sweet chestnut trees in the eastern United States began to be infected by a killing disease caused by a fungus, *Endothia parasitica*, that came to be known as the chestnut blight (Pl. 4). It was brought from Asia on nursery plants. In 1913 the parasitic fungus was found on its natural host in Asia, where it does no harm to the chestnuts. But the eastern American

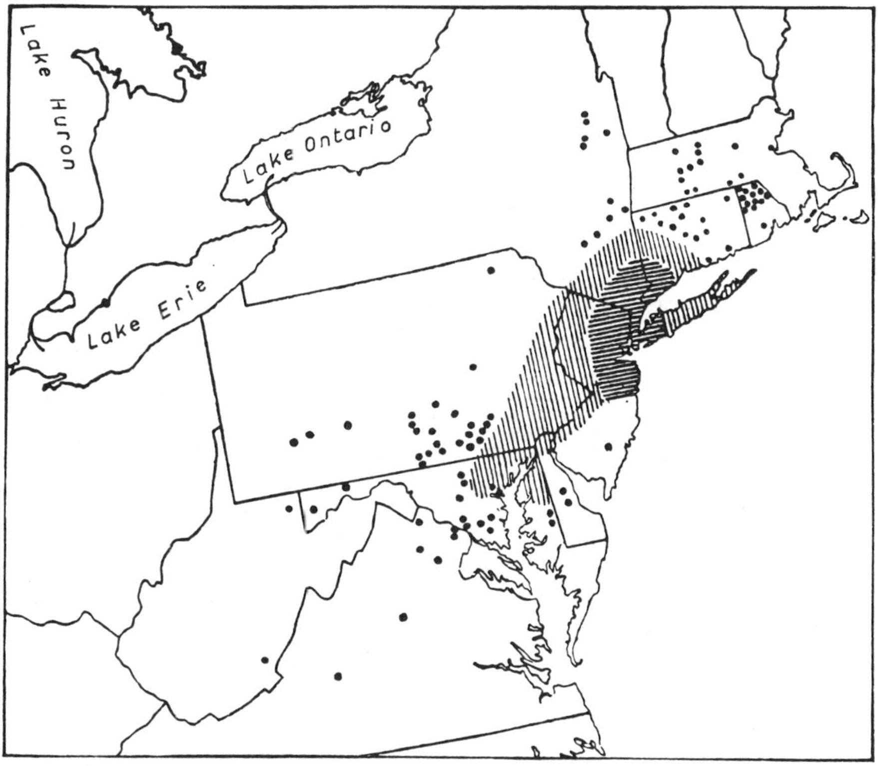

FIG. 4. Spread of the Asiatic chestnut blight, *Endothia parasitica*, to American chestnuts, *Castanea dentata*, in ten states. *Horizontal hatching:* majority of trees already dead; *vertical hatching:* complete infection generally; *dots:* isolated infections, many of which had been eradicated. (From H. Metcalft and J. F. Collins, 1911.)

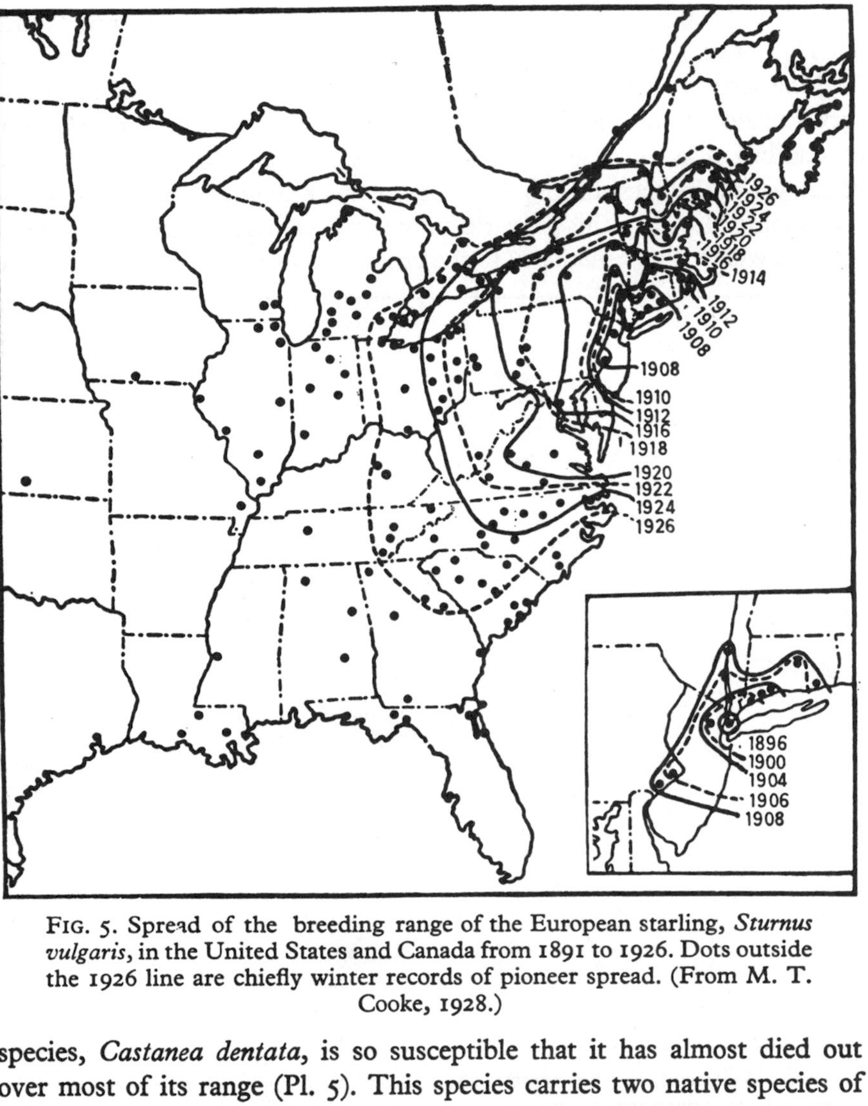

FIG. 5. Spread of the breeding range of the European starling, *Sturnus vulgaris*, in the United States and Canada from 1891 to 1926. Dots outside the 1926 line are chiefly winter records of pioneer spread. (From M. T. Cooke, 1928.)

species, *Castanea dentata*, is so susceptible that it has almost died out over most of its range (Pl. 5). This species carries two native species of *Endothia* that do not harm it, occurring also harmlessly on some other trees like oak; one of these two species also comes on the chestnut, *C. sativa*, in Europe.[27] As the map shows (Fig. 4), even by 1911 the outbreak, being

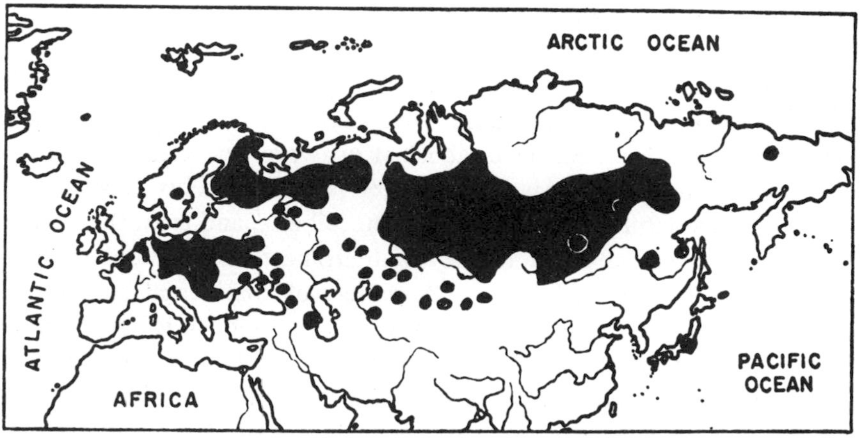

FIG. 6. Distribution of the North American muskrat, *Ondatra zibethica*, in Europe and Asia. (From A. De Vos, R. H. Manville and R. G. Van Gelder, 1956.)

through wind-borne spores, had spread to at least ten states, and the losses were calculated to be at least twenty-five million dollars up to that date.[23] In 1926 it was still spreading southwards, and by 1950 most of the chestnuts were dead except in the extreme south; and it is now on the Pacific coast too. So far, the only answer to the invasion has been to introduce the Chinese chestnut, *C. mollissima*, which is highly though not completely immune through having evolved into the same sort of balance with its parasite,[31] as had the American trees with theirs; much as the big game animals of Africa can support trypanosomes in their blood that kill the introduced domestic animals like cattle and horses. The biological dislocation that occurs in this trypanosomiasis is the kind of thing that presumably would have happened also if the American chestnut had been introduced into Asia. The Chinese chestnut is immune both in Asia and America. Already by 1911 the European chestnuts grown in America had been found susceptible.[23] In 1938 the blight appeared in Italy where it has exploded fast and threatens the chestnut groves that there are grown in pure stands for harvesting the nuts; it has also reached Spain and will very likely reach Britain in the long or short run.[8] Unfortunately the Chinese chestnut will not flourish in Italy, and hopes are placed solely on the eventual breeding of a resistant variety of hybrid.

The third example is the European starling, *Sturnus vulgaris*, which has spread over the United States and Canada within a period of sixty years. (It has also become established in two other continents—South Africa and Australia, as well as in New Zealand.) This subspecies of starling has a natural range extending into Siberia, and from the north of Norway and Russia down to the Mediterranean. We should therefore expect it to be adaptable to a wide variety of continental habitats and climate. Nevertheless, the first few attempts to establish it in the United States were unsuccessful. Then from a stock of about eighty birds put into Central Park, New York, several pairs began to breed in 1891. After this the increase and spread went on steadily, apart from a severe mortality in the very cold winter of 1917–18. But up to 1916 the populations had not established beyond the Allegheny Mountains. Cooke's map of the position up to the year 1926 (Fig. 5) shows how the breeding range had extended concentrically, with outlying records of non-breeding birds far beyond the outer breeding limits, which had moved beyond the Alleghenies but nowhere westward of a line running about southwards from Lake Michigan.[3] By 1954 the process was nearly reaching its end, and the starling was to be found, at any rate on migration outside its breeding season, almost all over the United States, though it was not fully entrenched yet in parts of the West coast states. It was penetrating northern Mexico during migration, and in 1953 one starling was seen in Alaska.[17] This was an ecological explosion indeed, starting from a few pairs breeding in a city park; just as the spread of the North American muskrat, *Ondatra zibethica* (Pl. 8), over Europe was started from only five individuals kept by a landowner in Czechoslovakia in 1905 (Fig. 7). The muskrat now inhabits Europe in many millions, and its range has been augmented by subsidiary introductions for fur-breeding, with subsequent establishment of new centres of escaped animals and their progeny (Fig. 8). Since 1922, over 200 transplantations of muskrats have been started in Finland, some originally from Czechoslovakia in 1922, and the annual catch is now between 100,000 and 240,000.[1] Independent Soviet introductions have also made the muskrat an important fur animal in

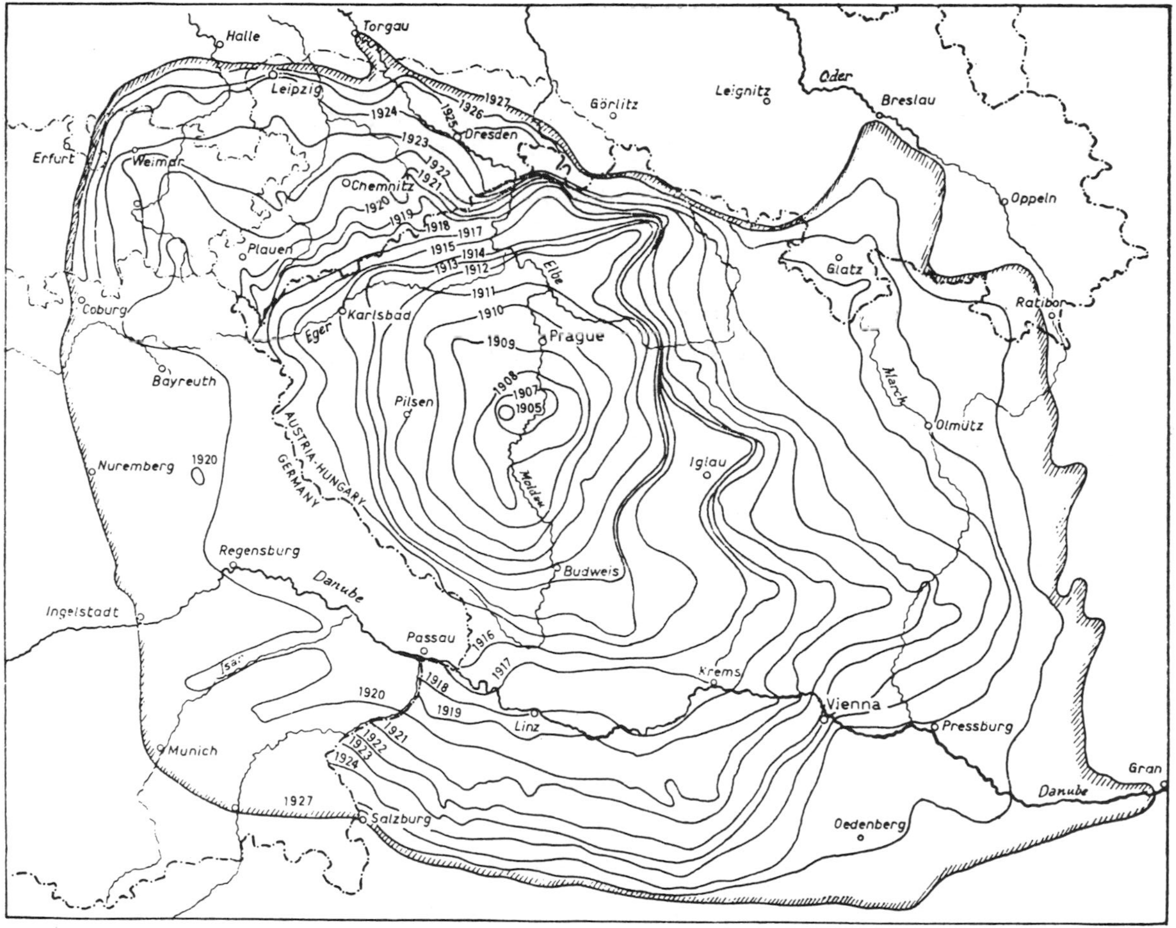

FIG. 7. Spread of the muskrat, *Ondatra zibethica*, up to 1927, from five individuals introduced into Bohemia in 1905. (After a coloured map in I. Ulbrich, 1930.)

most of the great river systems of Siberia and northern Russia, as well as in Kazakstan.[18] In zoogeographical terminology, a purely Palaearctic species (the starling) and a purely Nearctic species (the muskrat) have both become Holarctic within half a century (Fig. 6).

The fifth example is a plant that has changed part of our landscape—the tall strong-growing cord-grass or rice-grass, *Spartina townsendii*, that has colonized many stretches of our tidal mud-flats.[14] It is a natural hybrid between a native English species, *S. maritima*, and an American species, *S. alterniflora*, the latter brought over and established on our South coast in the early years of the nineteenth century. The strong hybrid, which breeds true, was first seen in Southampton Water in 1870, and for thirty years was not particularly fast-spreading. But during the present century it has occupied great areas on the Channel coast, not only in England but also on the North of France (Pls. 6–7). It has also been planted in some other places in England, and has been introduced into North and South America, Australia and New Zealand. The original American parent has largely been suppressed or driven out by the hybrid form. Here is a peculiar result of the spread of a species by man: the creation of a new polyploid hybrid species, from parents of Nearctic and Palaearctic range, which then becomes almost cosmopolitan by further human introduction. And it is on the whole a rather useful plant, because it stabilizes previously bare and mobile mud between tide-marks, on which often no other vascular plant could grow, helps to form new land and often in the first instance provides salt-marsh grazing. Its effects upon the coastal pattern are, however, not yet fully understood by physiographers and plant ecologists; but Tansley remarks that 'no other species of salt-marsh plant, in north-western Europe at least, has anything like so rapid and so great an influence in gaining land from the sea'.[29]

Changes of similar magnitude have been taking place in fresh-water lakes and rivers, as a result of the spread of foreign species. The sixth example given here concerns the sea lamprey, *Petromyzon marinus*, in the Great Lakes region of North America.[7] This creature is a North Atlantic river-running species, mainly living in the sea, and spawning in streams. But in the past it established itself naturally in Lake Ontario, as well as in some small lakes in New York State. But Niagara Falls formed an insurmountable barrier to further penetration into the inner Great Lakes. In 1829 the Welland Ship Canal was completed, providing a by-pass into Lake Erie. But it was a further hundred years or so before any sea lampreys were observed in that lake. Then the invasion went with explosive violence. By 1930 lampreys had reached the St Clair River, and by 1937 through it to Lake Huron and Lake Michigan, where they began to establish spawning runs in the streams flowing to these lakes. In 1946 they were in Lake Superior. Meanwhile the lampreys were attacking fish, especially the lake trout, *Salvelinus namaycush*, a species of great commercial importance. The sea lamprey is a combination of hunting predator and ectoparasite: it hangs on to a fish, secretes an anticoagulant and lytic fluid into the wound, and rasps and sucks the flesh and juices until the fish is dead, which may be after a few hours or as long as a week (Pl. 9). The numbers of lake trout caught had always fluctuated to some extent, and the statistics of the fishery since 1889 have been thoroughly analysed. But never before the recent catastrophe had the catch collapsed so rapidly: in ten years after the lamprey invasion began to take effect, the numbers of lake trout taken in the American waters of Lake Huron and Lake Michigan fell from 8,600,000 lb. to only 26,000 lb. On the Canadian side things were little better.[12] This was not caused by change in fishing pressure. Other species besides the lake trout have also been hard hit. Among these are the lake whitefish, burbot, and suckers, all of which declined in numbers. So, the making of a ship canal to give an outlet for produce from the Middle West has brought about a disaster to the Great Lakes fisheries over a century later. But in Lake Erie lampreys did not multiply, partly because there are not many lake trout there, but probably also because the streams are not right for spawning in.[19]

The seventh example is the Chinese mitten crab, *Eriocheir sinensis*, a two-ounce crab that gets its name from the extraordinary bristly claws that make it look as if it was wearing dark fur mittens (Pl. 10). At home it inhabits the rivers of North China, and it has been found over 800 miles up the Yang Tse Kiang. However, it breeds only in the brackish estuaries, performing considerable migrations down-stream for the purpose. The females don't move so far away from the sea as the males, and they can lay up to a million eggs in a season, which hatch into a planktonic

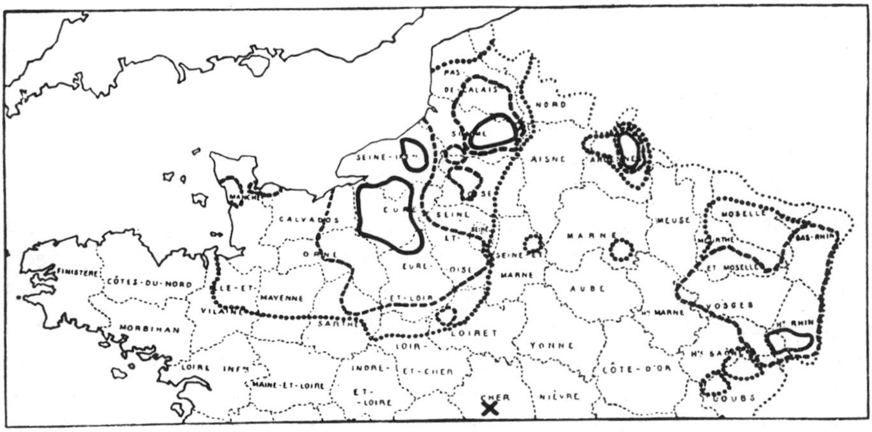

FIG. 8. Spread of the muskrat, *Ondatra zibethica*, in France. *Unbroken line*, 1932; *dashed line*, 1951; *dotted line*, 1954. *Cross*, one muskrat caught, extent of occupation unknown. (From J. Dorst and J. Giban, 1954.)

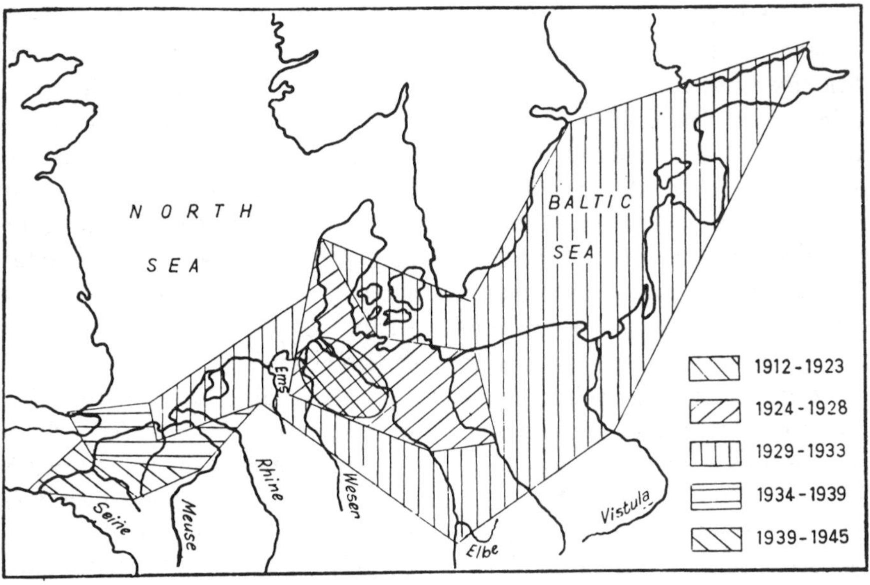

FIG. 9. Zones of spread of the Chinese mitten crab, *Eriocheir sinensis*, in Europe, 1912–43. (From H. Hoestlandt, 1945.)

larva (Pl. 11) whose later Megalopa stage migrates up-river again.[26] It is not really known how they got from East to West; they were first seen in the River Weser in 1912. The most likely explanation is that the young stages got into the tanks of a steamer and managed to get out again on arrival. Two large specimens were actually found in the sea-water ballast tanks of a German steamer in 1932, having, it is thought, got in locally from Hamburg Harbour. But these tanks are normally well screened. In the last forty-five years, mitten crabs have colonized other European rivers from the Baltic to the Seine (Fig. 9). Those that invaded the Elbe have arrived as far as Prague, like Karel Čapek's newts. This crab has not yet taken hold in Britain, though it may very likely do so some day, as one was caught alive in a water-screen of the Metropolitan Water Board at Chelsea in 1935.

These seven examples alone illustrate what man has done in deliberate and accidental introductions, especially across the oceans. Between them all they cover the waters of sea, estuary, river, and lake; the shores of sea and estuary; tropical and temperate forest country, farm land, and towns. In the eighteenth century there were few ocean-going vessels of more than 300 tons. Today there are thousands. A Government map made for one day, 7 March 1936, shows the position of every British Empire ocean-going vessel all over the world. There are 1,462 at sea and 852 in port; and this map does not include purely coasting vessels. Some idea of what this can mean for the spread of animals can be got from the results of an ecological survey done by Myers, a noted tropical entomologist, while travelling on a Rangoon rice ship from Trinidad to Manila in 1929. He amused himself by making a list of every kind of animal on board, from cockroaches and rice beetles to fleas and pet animals.[25] Altogether he found forty-one species of these travellers, mostly insects. And when he unpacked his clothes in the hotel in Manila, he saw some beetles walk out of them. They were *Tribolium castaneum*, a well-known pest of stored flour and grains, which was one of the species living among the rice on the ship.

A hundred years of faster and bigger transport has kept up and intensified this bombardment of every country by foreign species, brought accidentally or on purpose, by vessel and by air, and also overland from beard jutting out: 'We have been privileged to be present at one of the typical decisive battles of history—the battles which have determined the fate of the world.' But how will it be decisive? Will it be a Lost World? These are questions that ecologists ought to try to answer.

1. Dr Karl Meyer explaining methods of field survey for sylvatic plague to public health students, near San Francisco, 1938. (Photo C. S. Elton.)

2. Dissecting ground squirrels to obtain organs for plague testing, near San Francisco, 1938. (Photo C. S. Elton.)

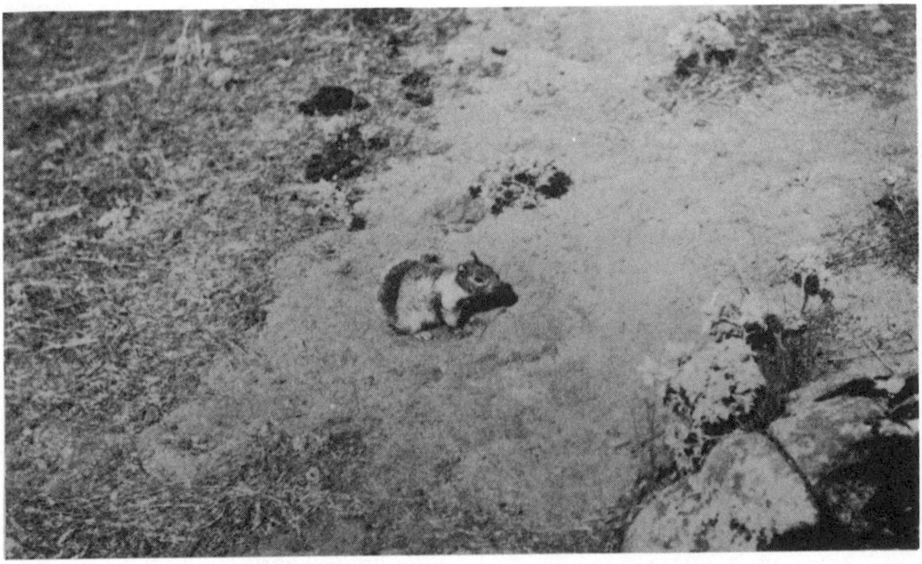

3. Ground squirrel, *Citellus beecheyi*: one of the wild hosts of plague in California. This one was in a plague-free part of the Sierra Nevada. (Photo C. S. Elton, 1938.)

4. White or buff-coloured mycelial fans of the chestnut blight, *Endothia parasitica*, seen after peeling bark off a diseased American chestnut. (From G. F. Gravatt and R. P. Marshall, 1926.)

5. An American chestnut, *Castanea denata*, almost killed by blight, *Endothia parasitica*, introduced from Asia. The new sprouts from the trunk would in turn become infected and die. (From G. F. Gravatt and R. P. Marshall, 1926.)

6 & 7. Muddy salt-marsh dominated by *Spartina townsendii*, at the head of a tidal inlet on the Sussex coast. Just off the lower area there was formerly a tidal water-mill, which has disappeared through the growth of the *Spartina* marsh in the last fifty years. (Photo M. J. Thornton, 1957.)

8. Muskrats, *Ondatra zibethica*, in their natural habitat in Montezuma marshes, New York State. (By E. J. Sawyer, in C. E. Johnson, 1925.)

9. Young sea lampreys, *Petromyzon marinus*, attacking brook trout in an aquarium. (From R. E. Lennon, 1954.)

10 & 11. Above: Male mitten-crab, *Eriocheir sinensis*, with claws raised. Below: Young planktonic stage. (From N. Peters and A. Panning, 1933.)

12. A family of coypus, *Myocastor coypus*, at home in South America. (From coloured painting by C. C. Wiedner in A. Cabrera and J. Yepes, 1940.)

13. Habitat of the South American coypu, *Myocastor coypus*, in the channel of an East Anglian broad. Cover and food are given by the luxuriant fen vegetation, in this photograph chiefly the reed-grass, *Glyceria maxima*. (Photo C. S. Elton, 1957.)

CHAPTER TWO

Wallace's Realms: the Archipelago of Continents

It is one of the first themes of this book that if we are to understand what is likely to happen to ecological balance in the world, we need to examine the past as well as the future. If during the last 100 million years the flora and fauna of the world had been able to develop in such a way that every organism had a good chance of spreading to all parts of the globe that its characteristics could tolerate, so that there was only one species for each kind of ecological situation, the potentialities of future change under the impact of man's activities would be different. They would be far less, though still considerable, because man has altered habitats as well as moving species around like chessmen. In the kind of world described, where there were no barriers to spread, we should have mostly pan-tropical and pan-temperate species (we do mostly have pan-arctic ones as it is), bipolar forms, continental species reaching every island, fresh-water species moving freely to all isolated waters, marine animals also girdling the world and reaching northern and southern hemispheres. The rabbit might already have been in Australia, the coypu in East Anglia, the mitten crab in the Elbe, and the giant snail in the Mariana Islands. That is to say, they would if they had evolved successfully in face of rival lines.

Quite a large number of species are able to achieve a world-wide distribution as it is, either because the ecological barriers that hold in others are not barriers to them, or because, which is partly the same thing, they have exceptionally good powers of dispersal. *Calanus finmarchicus*, the most abundant copepod crustacean in the plankton of North Atlantic seas, can get to the Indian Ocean near Madagascar because the cold northern currents dive downwards and travel below the warmer surface waters of the tropical seas: in this way the copepod has crossed the Line, or rather under it. Many birds migrate across the world—the Arctic tern can go from Arctic to Antarctic, the golden plover right down and up the Americas, the swallow to and from Europe and South Africa, and there are flight-lines of waterfowl between Australia and Japan. Microscopic forms whose eggs or dried bodies float on wind or get caught in birds' feathers are often world-wide in distribution. Such are many Protozoa, rotifers and waterfleas—not to mention many seeds of plants. Besides these, there are a good many mobile forms that have gradually covered the world, in spite of sea, mountain, and desert. One of the best known of these is the barn owl.

But a great many other plants and animals never had the opportunity of ranging over the whole world. The meaning of Wallace's Realms is that these became cooped up, as it were, in various regions for long enough to change profoundly and leave a permanent mark on the composition of flora and fauna. To this were added two other processes that increase the complexity of the distribution pattern as we see it now. The first is that groups that were formerly very widespread have retreated and may be found, say, only in one continent or island or lake. The second is that after the long periods of isolation in Tertiary times that created Wallace's Realms, there was some remingling of faunas before man came on the scene to carry the process abruptly further. For an example of the first process, redwood trees (*Sequoia*) used to grow right across North America and in Eurasia, though now they are confined to California and Oregon. For one of the second process, we know also from fossils that tapirs evolved inside an isolated North America, but they have spread to Central and South America and South-east Asia, where they still live, though extinct in their original home. These are the three causes of diversification: the breaking up of an ancient cosmopolitan pattern, the evolution of regional groups, and the partial randomization of these regional groups over the world again. If this randomization had been complete, we would not be able to detect Wallace's Realms at the present day, or if they were visible they would mean something very different.

This book is not meant to supply a critique of zoogeography, but only tries to pick out some of the simpler realities from a vast field in which the subjects are half hidden from the ordinary inquirer by deep screes of uncritical 'facts', dubious theories, and information that has never been used at all for a zoogeographical purpose. In contemplating this enormous and indigestible subject, one almost envies Wallace himself, who could at any rate lie in his camp with fever and think of the rather elementary proposition of the struggle for existence and natural selection; a proposition so elementary that only one other man had ever fully worked it out before! His own modest assessment of the Realms, or Regions as he called them himself, is worth giving here: 'Our object is to represent as nearly as possible the main features of the distribution of existing animals, not those of any or all past geological epochs. Should we ever obtain sufficient information as to the geography and biology of the earth at past epochs, we might indeed determine approximately what were the Pliocene or Miocene or Eocene zoological regions; but any attempt to exhibit all these in combination with those of our own period, must lead to confusion.'[57]

Before describing the Realms it is necessary to look at the state of distribution in the world before the Cretaceous Period. There is fairly general agreement among geologists that all through fossil history from Palaeozoic times until the present, and in spite of many changes in detail of the coastlines in the world, there have always been some bodies of land corresponding to the present continents. Great parts of these land masses have never been under the sea at all. Marine life began to press on to the land and fresh waters perhaps in Silurian, certainly in Devonian times, say (with a fairly big error in estimates) about 315 million years ago, or more. During the next 230 million years, up to the end of the Cretaceous Period, there was never a time when anything at all closely resembling Wallace's Realms could be discerned from the fossil picture. Plenty of regional differences from time to time, especially between Northern and Southern Hemispheres; climatic changes like the Permian Ice Ages, dry and wet periods, greater or lesser land surface. But as each new major group evolved, it is quite plain that it eventually spread round the whole world, to all or nearly all the areas that are at present continents, and to some that are now islands, like Madagascar. This is what can be called the period of cosmopolitan distribution. When we come to the Mesozoic Age, and particularly to the middle and later parts comprising the Jurassic and

Cretaceous Periods, there is strong evidence that the world's climate was either more uniform or at any rate warmer than it is now. One has to accept very great changes of some kind, to account for luxuriant forests in Greenland and Spitsbergen and an Arctic Ocean filled with abundant ammonites and other marine forms.[33]

At some time during or not very long after the Cretaceous Period, according to the part of the world concerned, there was considerable transgression of the sea on to land, and this eventually broke land connexions in many parts of the world. It happened in the Panama Isthmus, in Bering Strait, with the connexions to Australia, and elsewhere. The timing of this gigantic series of events, which made the continents into an archipelago, was a particularly important accident of history. A great many groups of plants and animals had already evolved far towards their present state before it happened. By the end of the Cretaceous many modern genera of trees like oak, poplar, beech, sycamore, magnolia, laurel, pine, spruce, and cedar already existed. Also most of the present-day families, a good many genera, and even some species of insects were evolved. But the mammals and birds and fresh-water fish were just, as it were, poised on the edge of a tremendous bout of evolution; when the continents were separated each one had its quota of early forms of mammals and birds, and some of them also of fresh-water fish which then went along quite separate lines as the different Realms were cut off.

Wallace divided the world into six Regions, using names adapted from the continents or of a classical form. To this task he brought a wonderfully rich experience of some of the most exciting facts of zoogeography, from his personal explorations of the Rivers Amazon and Rio Negro, in 1848–52, and of the Malay Archipelago in 1854–62. His first ideas about geographical distribution were written while he was still in the East, for they were published in 1860. He returned later, having collected altogether 125,660 specimens of animals and discovered what came to be known as Wallace's Line. His six regions can be quite simply defined, though his own account of the features of these realms filled a two volume book of 1,110 pages, and can scarcely be summarized here. The Neotropica Region covers Southern and Central America up to a line in Mexico together with the West Indies (Pl. 14). North America, its Arctic islands and Greenland form the Nearctic Region (Pl. 15). The Palaearctic Region is Europe and part of Asia, running across from Britain to Japan (Pl. 16). Within Asia he separated the Oriental Region (Pl. 19), formed of the Indian Peninsula, the Far East with southern China and Formosa, also the Philippine Islands, and all Malayan islands west of Celebes and Lombok. The northern boundary of the Oriental Region is mainly given by the Himalaya and great mountain ranges east of it, supported also by the desert barriers north of them. The Ethiopian Region (Pl. 17) is Africa south of the Sahara, and Madagascar. The Australasian Region (Pl. 18) takes in the islands east of Borneo, Java, and Bali, with New Guinea, and Australia; also New Zealand and the Pacific islands. The usefulness of these regions, or realms, has been truly proved in the century since they were proposed, though endless discussions of details about their limits and subdivisions and composition and history have gone on. Wallace himself thought that New Zealand was an anomaly on its own. And I would say that the vast Eastern Pacific beyond the Australian continental arc, with its Milky Way of islands mostly with exiguous numbers of species (even in Hawaii), deserves a totally separate treatment. There are, in fact, seven great 'realms of life'.

It is possible to give a brief picture of the history of some of these realms, because they are either quite well understood, as with the Americas and Europe, or there is a comparative blank in fossil records as in Australia and Africa. For Asia the picture is patchy, and history rests chiefly on other inferences. It is convenient to start with the Neotropical Region. When the *Beagle* sailed from Bahia Blanca in the summer of 1833, Charles Darwin stayed on shore to make an overland journey to Buenos Aires.[36] It was on his way there that he came across a deposit of large fossil bones that included four kinds of huge extinct ground sloths, two other kinds of edentates, a horse, a toxodont, and what he thought to be the tooth of an animal in the group now known as Litopterna. Later on, in January of the following year, he found half a skeleton, 'full as large as a camel', embedded in the red Pleistocene mud of a gravel plain on the pampas of Patagonia. This was *Macrauchenia*, the last survivor of the Litopterna, of which Scott remarked that it 'must have been one of the most grotesque members of this assemblage of nightmares, as it would have seemed to our eyes'.[51]

The Litopterna are one of several strange groups of hoofed animals, of which the Toxodonta are another, that never went outside the Neotropical Region, evolving and dying out there during the Tertiary isolation. Geologists have direct evidence of the complete break through of ocean straits between North and South America from the middle of the Eocene to the middle of the Pliocene—an enormous length of time (Fig. 10). (To

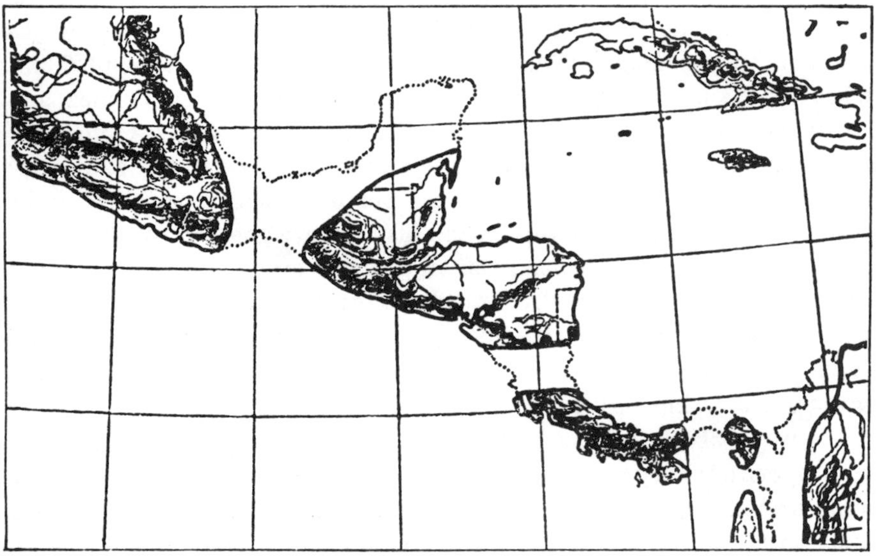

FIG. 10. The broken Isthmus of Panama in Tertiary times, according to the conclusions of geologists. These straits were not in existence individually for the whole period, but between them spanned the middle Eocene to middle Pliocene. (From E. Mayr, 1946.)

avoid confusion in the reader's mind, the order of the Tertiary periods in geology is given here: Eocene, Oligocene, Miocene, Pliocene; followed by the Quaternary divisions of Pleistocene (the Ice Age) and Holocene (since the Ice Age).) The ancestors of the placental mammals that gave rise to the Litopterna and other endemic groups of South America appear to have managed to enter the continent in the comparatively short period called Paleocene, between the end of the Cretaceous Period and the Eocene, and also in the early Eocene. Marsupials had also been there before the break, and evolved thereafter abundantly into forms that still survive in some places as small land and water animals. The carnivores were marsupials, one of which looked remarkably like the sabre-toothed tiger. There was a riotous evolution of extraordinary edentates, which survive as anteaters, sloths, and armadillos, but in earlier times included also giant armadillos and giant sloths of the strangest kind. There were and still are many unique groups of rodents—Simpson suggests that the ancestral rodents and South American monkeys may originally have got across the straits by 'island hopping'.[52] It is indeed well known that tropical rivers carry to sea rafts of vegetation sometimes with animals on them. For instance, a green monkey was noticed on floating timber near Java in 1883,[39] and a fer de lance snake arrived on flotsam on the coast of Peru after changes in the force of the Equatorial counter current.[45] Dammerman cites other examples of the kind.[35] But one could not get a toxodont or a litoptern, or a chinchilla the size of a rhinoceros, over the waters on a small raft of vegetation.

This hard evidence from geology and fossils, as well as present peculiarities of the fauna, prove that there was a long isolation. No doubt if birds were more successfully preserved as fossils, we should be able to follow a very similar development of the many families of birds that are now endemic in the Neotropical Region—such as rheas, screamers, oil-birds, toucans, honey-creepers, wood-hewers, puff-birds and others. And the isolation has left a profound mark on the composition of the whole flora and fauna. Tertiary fossils from the Argentine and Chile include no genera of North American plants except ginkgo. Most of the northern genera and species are still absent from Chile though they grow if introduced.[34] But it did not continue after the middle Pliocene, when the renewal of Panama Isthmus allowed intermixing between the faunas of north and south. Many of the creatures we know as South American arrived from the north after that time, having evolved elsewhere.

This assemblage of Pliocene invaders included the tapirs, llamas, peccaries, deer, foxes and dogs, cats, otters, bears, raccoons, and skunks, some of which survive still; and a very interesting company of mastodons, horses, American antelopes, voles, and the real sabre-toothed tiger, that did not survive to recent times, though man has brought in horses again.

It is an absolute historical fact that both the Pliocene invaders and the originally evolved inhabitants of the Neotropical and the Nearctic Regions underwent extraordinary casualties when the two faunas had been partly brought together after their long isolation from each other. Could it not be that this intermingling of species that had not evolved into ecological balance led to dislocations as catastrophic as the entry of the sea lamprey into the inner Great Lakes, or the spread of the Asiatic chestnut fungus in America? Of course the scale of time is totally different—one in millions and the other in decades; but the same principle could operate in both.

Turning to North America, there is an even more wonderful fossil story that has grown since the days of Wallace, and in some respects a similar development of special faunas.[51] There are the undoubted signs of long isolation yet combined with equally undoubted invasions that probably came in from Eastern Asia over the region of Bering Strait. On the one hand there are families of mammals that evolved wholly inside North America and have never been found outside it. Such are the subterranean rodents called pocket gophers, which are abundant at the present day; the pig-like oreodonts, that became extinct a long time ago; and the camels and tapirs, evolved there but now only found in other countries to which they managed to travel in Pliocene or Pleistocene times. On the other hand, the invaders from the Palaearctic Region, such as deer and members of the order of elephants that arrived in the Miocene. Yet no giraffes or ostriches ever made the crossing from Asia. The explanation seems to be that the land bridge over Bering Sea, in the words of De Chardin, 'was never a very broad or comfortable one. Like a constricted channel, it yielded only to lucky strokes, or to swift and adaptable animals, or to heavy biological pressures.'[37] Simpson, from a review of the fossil records, decided that 'Faunal resemblance between Eurasia and North America has been much lower at all times since the early Eocene than it is at present. It was especially low in the middle to late Eocene and late Oligocene to early Miocene. At those times, at least, the concept of a Holarctic Region is not applicable. In fact it hardly seems to apply at any times except the early Eocene and the present.'[52]

Some of the Neotropical animals also managed to invade North America over the new isthmus. The porcupine, successfully established as a forest animal right up into Canada and Alaska, belongs to one of the fourteen families of rodents that evolved in South America, where there are other forest porcupine species at the present day. A species of capybara, a large aquatic rodent also belonging to one of these endemic families, reached and occupied the southern parts of North America in the glacial period, but has since died out, as have also the great ground sloths and glyptodonts ('giant armadillos') that spread quite widely northwards from their original centre of evolution in the south. In 1941 a bone from the foot of a ground sloth was found in what Stock calls 'the frozen muck of Alaska'. This ground sloth was in a glacial stratum that has, in Alaska, produced the bones of saiga antelope, bison, and woolly mammoth (genera derived from the Palaearctic Region), as well as others like a large camel evolved within North America.[53]

In a short sketch one can only indicate that the history of Wallace's Realms really has been emerging like a photograph in a slow developer: the evidence is there, it is no longer just a theory that these colossal separate nature reserves of Tertiary times existed, there was an archipelago of continents for part of that time. Man is carrying on and accelerating an interchange of species that was going on some fifteen million years ago when some of the continents were joined again. And it is solid proof of the efficacy of the larger physical and ecological barriers that such realms of life still retain the strong impression of independent evolution taking place within what were often similar kinds of ecosystems like forest, desert, grassland, lake, and river. One wonders whether it is just a coincidence that the erasure of earlier differences by mutual migrations shows most of all in the simpler ecosystems of the Arctic tundra, whose fauna and flora are in so many respects circumpolar; that the next greatest resemblance is in the Boreal forests of Canada, Alaska, Kamchatka, Siberia, and north Europe;[54] and that as you move southward in North America or Eurasia these resemblances diminish, until in the tropical forests of Central and South America, Africa, and the Oriental Region, the rich accumulation of species shows most strongly of all the character of its past.

The Palaearctic, Oriental, and Ethiopian Regions were most strongly influenced by three features: the Tethys Sea, the Tertiary mountain-building movements, and the glaciations. The Tethys Sea girdled the world from Atlantic to Indo-Pacific. It was a very ancient sea, shallower than the main oceans, a sort of continental sea separating Eurasia from Africa and India but giving a highway for warm-water marine life between the two great oceans. At the present day it survives as the Mediterranean and Black Sea, with outlying relics in the Caspian Sea and Lake Aral; in the Red Sea and the Indo-Pacific area. It is often marked fairly imaginatively on broad geological maps of the world as a permanent barrier between the northern and southern continents throughout a large stretch of the Palaeozoic and Mesozoic Ages, i.e. up to the end of the Cretaceous Period at least. But the former cosmopolitan dispersal of so many land plants and land and fresh-water animals up to that date makes it impossible to believe that the Tethys Sea was never bridged by land; indeed, it must have been from time to time.

In the course of the Tertiary Period this enormous trough or geological syncline was partly heaved up into equally enormous mountain ranges from the Pyrenees to the Himalaya and beyond into South-east Asia and China. This mountain-building process was well advanced by the Miocene Period, and formed the northern barricade of the Oriental Region; and because it runs very roughly in a broad belt of equal latitudes (until it turns south at the extreme eastern end), it marks also a rather sharp line between tropic and temperate zones. That is to say, the Oriental Region is a relatively late realm, marked both by barriers to dispersal in the north, and by its generally tropical stamp. Had there never been these mountains and the deserts behind them, the region would now just be the tropical belt of the Palaearctic Region—but it would have retained a less rich museum of special forms from the past. In Pliocene times and perhaps later also, the nearer large islands of the Malay Archipelago were attached to the mainland: they are relatively modern continental islands, much younger than Madagascar or New Zealand, but older than Great Britain.

The raising up of the Tethys sea-bottom also cut the main channel between the eastern and western oceans in the Middle East. This in turn meant a broad land junction between the Palaearctic and Ethiopian Regions. The evidence for this change comes in two ways, apart from the direct general implications of the rising mountain ranges. In the first place, the resemblance between the Mediterranean part of Tethys and the eastern seas begins to diminish sharply during the Miocene Period. Until the end of the Oligocene there was a very rich tropical fauna all through these seas, with coral reefs, *Nautilus* and king-crabs, which do not now occur in the west Atlantic, and *Nautilus* not in the Atlantic at all.[38] The change in fossils conclusively proves that the Tethys Sea was severed during the Miocene.

The second kind of evidence comes from the fossils of land animals that came to the Palaearctic Region from Africa during the Tertiary, or vice versa. It is believed that the hyraxes and elephants evolved within the African continent—though in the early Tertiary times North Africa would be on the south side of Tethys Sea, and therefore 'Ethiopian', whereas it is now separated from Wallace's Ethiopian Region by deserts and has a Palaearctic stamp. There is still too scanty a fossil record for Africa to enable us to draw up any exact time-table of these crossings: for example, there are no Pliocene remains.[40] But in the last decade very rich collections of Miocene fossils, including mammals, insects, and plants, have been found in East Africa, notably on islands in Lake Victoria—when these are fully published a large vacuum in knowledge will be filled.[41] Africa seems to have evolved a peculiar fauna in early Tertiary times, but probably not on the scale of North and South America. Madagascar is a museum preserving some of these forms, through having been cut or rifted off from the mainland before the great invasions that entered Africa during the Miocene and Pliocene. These invaders probably included the greater part of the big game animals (such as antelopes) that we ordinarily think of as being peculiarly African. It is known that the hippopotamus and giraffe appeared in Africa in the lower Pleistocene, invading almost certainly from Asia.[40]

The Pleistocene Ice Age, or rather series of glaciations, had three crushing effects upon the distribution of animals in Europe. The spreading ice of course erased life during its advance (except on projecting nunataks of mountain); it drove southwards the various zones of life against the simultaneous movement of glaciations on the various alps (themselves a result of Tethys history, as has been explained); and while the snowfall and the snow-line changed in the north and on the temperate mountains, there were parallel changes in rainfall in Africa—pluvial periods that alternated with long times of drought. A fourth indirect effect of glaciation was to withdraw so much water from the sea as to create many land connexions from islands to each other and the mainland, and also affect the growth of coral reefs in the tropics. Europe suffered far the greatest catastrophe and impoverishment, and its flora and fauna are still much poorer than those of Eastern Asia, where the calamity was comparatively local. Every year sees the enrichment of our gardens out of the wealth of Chinese vegetation lasting from Pliocene times. Were it not for the Ice Age, we should probably have wonderful mixed forests with wild magnolias and laurels and epiphytic orchids, such as Hooker described about 110 years ago in his travels through Sikkim. In China there are about 500 species of trees![42]

Although Australia is a very ancient continent, the presence of some of the large dinosaurs of the late Cretaceous Period and other facts of the kind, prove it until then to have been in touch with the rest of the world. No placental mammals except bats and rodents and dingos were there when it was discovered by white men, and there was a luxuriant evolution of marsupials—a kind of mammal that has everywhere else died out except in South America (where they are much diminished from their former status) and in North America where there is one sort of opossum. Another even more primitive group, the egg-laying monotremes (platypus and echidna) also has its last outpost in Australia and New Guinea. Such facts seem to prove that the Region was isolated from Asia by or before very early Tertiary times. Yet the land and submarine map of the Malay Archipelago does not at first sight explain how such an isolation could have been maintained for perhaps eighty million years. But in the summer of 1856 Wallace sailed from Bali to the next island eastwards, Lombok, and he wrote: 'The hills were covered with a dense scrubby bush of bamboos and prickly trees and shrubs, the plains were adorned with hundreds of noble palm-trees, and in many places with a luxuriant shrubby vegetation. Birds were plentiful and interesting, and I now saw for the first time many Australian forms that are quite absent from the islands westward. Small white cockatoos were abundant, and their loud screams, conspicuous white colour, and pretty yellow crests, rendered them a very important feature in the landscape. This is the most westerly point on the globe where any of the family are to be found. Some small honey-suckers of the genus Ptilotis, and the strange mound-maker (Megapodius gouldii), are also here first met with on the traveller's journey eastward.'[56]

This abrupt change in the pattern of faunas was elaborated by Wallace in 1860, and later T. H. Huxley gave it the name of Wallace's Line. Northwards this Line runs between Borneo and Celebes, along the deep Macassar Strait, a very ancient geological feature that seems to have prevented fresh-water fish from spreading eastwards. An American field zoologist, Raven, who spent some years in Borneo and Celebes collecting animals, subsequently mapped all the records there were of mammals for the whole Malay Archipelago and adjoining regions.[48] These maps show how each group has spread out from its headquarters in Asia or Australia, the number of species thinning out towards the meeting of the two faunas in a central zone formed of Celebes, the Lesser Sunda Islands, and the Moluccas. From Asia extend such groups as the carnivores, insectivores, squirrels, elephants, other ungulates like rhinoceros and tapir, and the primates; from Australia the various groups of marsupials, a selection from the richer mainland fauna. Only the bats, as one would expect, range everywhere. Three maps (Figs. 11–13) made by Rensch illustrate these distributions also for birds and other groups. The central zone is known to have had a very disturbed geological history, and is at this day full of large and partly active volcanoes. Wallace's Line marks its western edge. On the eastern side there is another, perhaps more arbitrary line, called Weber's Line, running west of the Moluccas and east of Timor, which also marks the beginning of a poorer fauna and a diminution of Australasian forms.[43] It mostly follows the fifty-fathom line. Geologists consider that there was land across in the Mesozoic Period, but even this has been questioned by some zoogeographers on the ground that Australia, so poor in true fresh-water fish, ought therefore to have retained a great many primitive Mesozoic fish at the present day. (If this is true, the great reptiles would have to have reached Australia by an Antarctic route from Patagonia.) The 600–800 mile belt of unstable land that subsequently cut off Australasia from Asia is the barrier that led to the development of another of Wallace's Realms. In the early part of the Tertiary Borneo and Java themselves were probably not yet above the sea so that the zone was wider

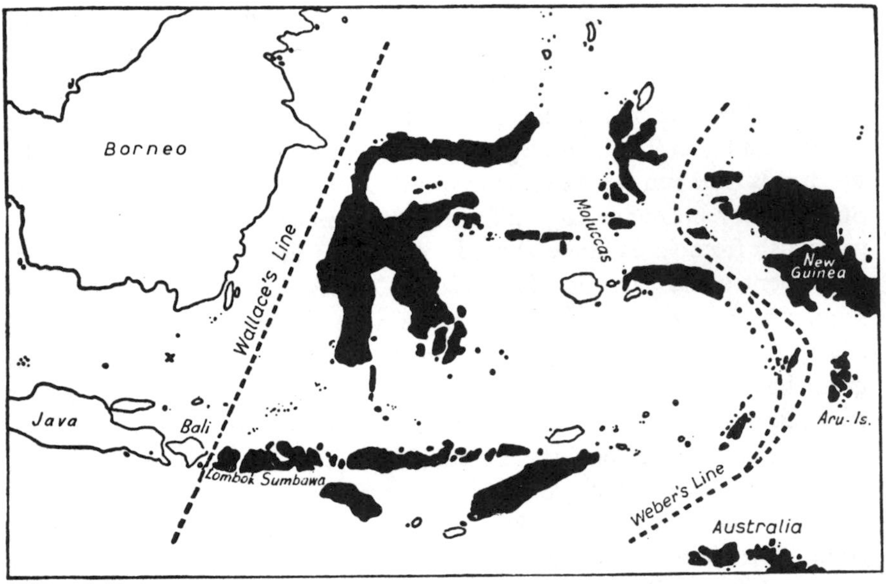

FIG. 11. Distribution of an Australasian family of birds, the cockatoos, westwards to Wallace's Line. The only record west of the Line is shown by a cross. (After B. Rensch, 1936.)

Courtesy: Gebrüder Borntraeger

then. But it might be suggested (which I have not seen done) that there could be another reason for the effectiveness of this barrier—the ecological dislocation of occasional arrival by invaders from either side onto islands with relatively incomplete communities.

The definition of the Australasian Region needs only to be rounded off by noting that the wonderful series of islands east of Australia—including New Zealand, New Caledonia, New Hebrides, the Solomons, and Fiji—are part of what Suess called 'the shattered remnants of a foundered continent'. Farther east the islands are all completely oceanic in origin, formed from volcanoes or coral reefs grown upon them. These oceanic islands never were joined to land, and their fauna and flora is accordingly poor and derived from stragglers accidentally arriving over long periods of time. We might really call this the Pacific Oceanic Region. Even though this Region has never had a continent, one can hardly leave out of con-

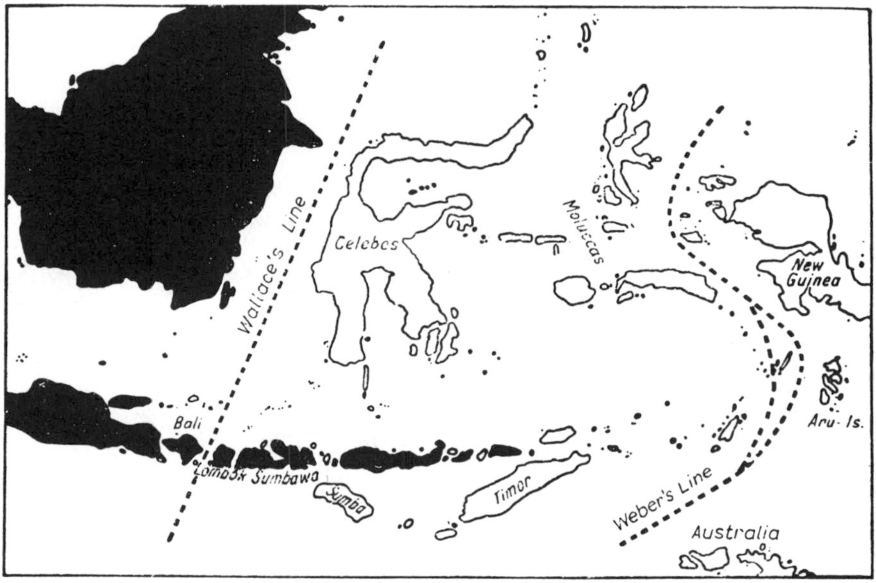

FIG. 12. Distribution of a woodpecker, *Dryobates moluccensis*, west of Wallace's Line but also in the smaller Sunda Islands east of it. (After B. Rensch, 1936.)

Courtesy: Gebrüder Borntraeger

sideration the island zoogeography of an ocean that is larger in area than all the continents and islands of the world combined! Ecologically, the modern invasions of these islands are among the most interesting, though lamentable events of modern times (Chapter 4).

The distribution of fresh-water fish gives a very remarkable proof, if any more were needed, of the timing and development of Wallace's Realms. True fresh-water fish, that is species that do not run to estuaries or the sea, are not likely to be dispersed across the sea except by rare accidents. Their slow dispersal depends very much on the changes of water systems, the capture of watersheds by rivers, the joining of river mouths by elevation of the land, and a certain amount of local 'lake- and river-hopping' by accidental dispersal through wind and birds and also early human agencies.[47] It happened that the vast group of modern fish called Ostariophysi, that includes most of the world's fresh-water species and about a

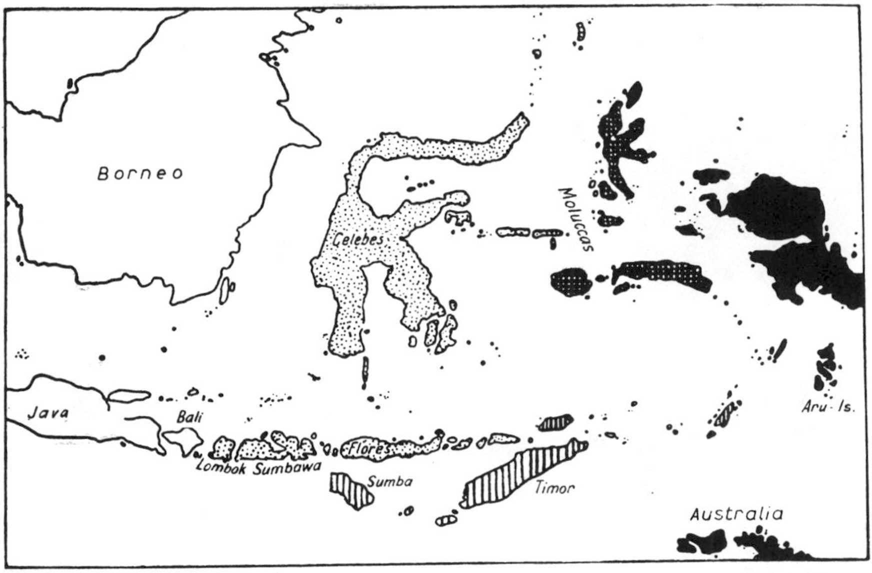

FIG. 13. A summary of the distribution of Australasian genera of mammals, birds, reptiles, amphibia, butterflies, and land snails, in the region of Malay Archipelago and the western New Guinea islands. The symbols give the percentage of such genera in various islands: *black:* 76–100; *cross-hatched:* 46–55; *wide vertical hatching:* 31–45; *close vertical hatching:* 11–30; *dots:* 0–10. (After B. Rensch, 1936.)
Courtesy: Gebrüder Borntraeger

quarter of all species of fish, was just in process of evolution at the end of Cretaceous times. They were evolving fast at the end of that Period and in the early Eocene. There are two groups of older fresh-water fish—the lung-fish and the Osteoglossidae—that have species still living round the world (lung-fish in Australia, Africa, and South America; osteoglossids in Northern Australia, Borneo, Sumatra, Malaya, the Upper Nile, West Africa, and South America). The lung-fish are a group that was present in Palaeozoic times, and the Australian kind was cosmopolitan in the whole Mesozoic Age. The Osteoglossidae have early forms in the Eocene of North America and in the Cretaceous and Eocene of England.[50] But the 30 families of the Ostariophysi are practically absent altogether from Australasia. 'Makassar Strait forms the most spectacular zoogeographical boundary to be found among the world's fresh-water fish faunas. To the west lies Borneo, teeming with 17 families and 300 or more species of primary fresh-water fishes. Only eighty-five miles to the east lies Celebes, with two solitary species of primary fresh-water fishes, both probably introduced by man.'[46]

One large section of the Ostariophysi, the catfish or Siluroidea, has a few species in Australia that are supposed to have become freshwater secondarily after an intermediate marine evolution. The Neotropical Region now has nine endemic families of catfish, the Nearctic one family except for a species in China, while there are also endemic families in the Ethiopian and Oriental Regions. So slow are fresh-water fish to become redistributed across renewed land junctions that they can almost be called 'living fossils', in so far as their present distribution is often one or more geological periods behind that of the more mobile mammals.

CHAPTER THREE

The Invasion of Continents

When contemplating the invasion of continents and islands and seas by plants and animals and their microscopic parasites, one's impression is of dislocation, unexpected consequences, an increase in the complexity of ecosystems already difficult enough to understand let alone control, and the piling up of new human difficulties. These difficulties have mounted especially in the last 150 years, and they have had to be met by means of a series of fairly hasty and temporary measures of relief that are only here and there supported by fundamental research on populations, or even a systematic record of events. Indeed it is easy to feel like Edward Gibbon, who wrote at the end of *The Decline and Fall of the Roman Empire*: 'The historian may applaud the importance and variety of his subject; but, while he is conscious of his own imperfections, he must often accuse the deficiency of his materials.' This is not, however, to criticize the biological workers who have had to grapple with an unending string of unforeseen emergencies with the scanty means at hand; and there are a certain number of remarkably fine and carefully compiled histories of invasions, notably by the various branches of the United States Department of Agriculture, who were the first to bring some sort of method and order into this field. In the present chapter it will only be possible to select a few examples, and these are not so much chosen for their economic or medical or veterinary importance, as to illustrate the ideas of this book, or because they happen to have good maps of the invasions of continents by foreign species.

No one really knows how many species have been spreading from their natural homes, but it must be tens of thousands, and of these some thousands have made a noticeable impact on human life: that is, they have caused the loss of life, or made it more expensive to live. If we look far enough ahead, the eventual state of the biological world will become not more complex but simpler—and poorer. Instead of six continental realms of life, with all their minor components of mountain tops, islands and fresh waters, separated by barriers to dispersal, there will be only one world, with the remaining wild species dispersed up to the limits set by their genetic characteristics, not to the narrower limits set by mechanical barriers as well. If we were to build six great tanks, fill them with water and connect them all to each other by narrow tubing blocked by taps; then fill these tanks with different mixtures of a hundred thousand different chemical substances in solution; then turn on each tap for a minute each day; the substances would slowly diffuse from one tank to another. If the tubes were narrow and thousands of miles long, the process would be very slow. It might take quite a long time before the whole system came into final equilibrium, and when this had happened a great many of the substances would have recombined and, as specific compounds, disappeared from the mixture, with new ones or substitutes from other tanks taking their places. The tanks are the continents, the tubes represent human transport along the lines of commerce; but it has not proved possible to turn off the taps completely, even though we might often wish to do so. And although there is a Law of the Conservation of Matter, there is no Law of the Conservation of Species.

One of the primary reasons for the spread and establishment of species has been quite simply the movement around the world by man of plants, especially those intentionally brought for crops or garden ornament or forestry. Fairchild, who was head of the United States Office of Plant Introduction, mentions casually in a travel book about the tropics that the work of this organization 'has resulted in the introduction of nearly 200,000 named species and varieties of plants from all over the world'.[78] This is a very solid contribution to the vegetation of nations! Just as trade followed the flag, so animals have followed the plants. For example, in the summer of 1916 about a dozen strange chafer beetles were noticed in a plant nursery in New Jersey. These were identified as *Popillia japonica* and called the Japanese beetle. From this centre the population grew rapidly outwards.[102] In the first year the beetles covered less than an acre. In the next seven years the areas inhabited increased as follows: 3, 7, 48, 103, 270, 733,

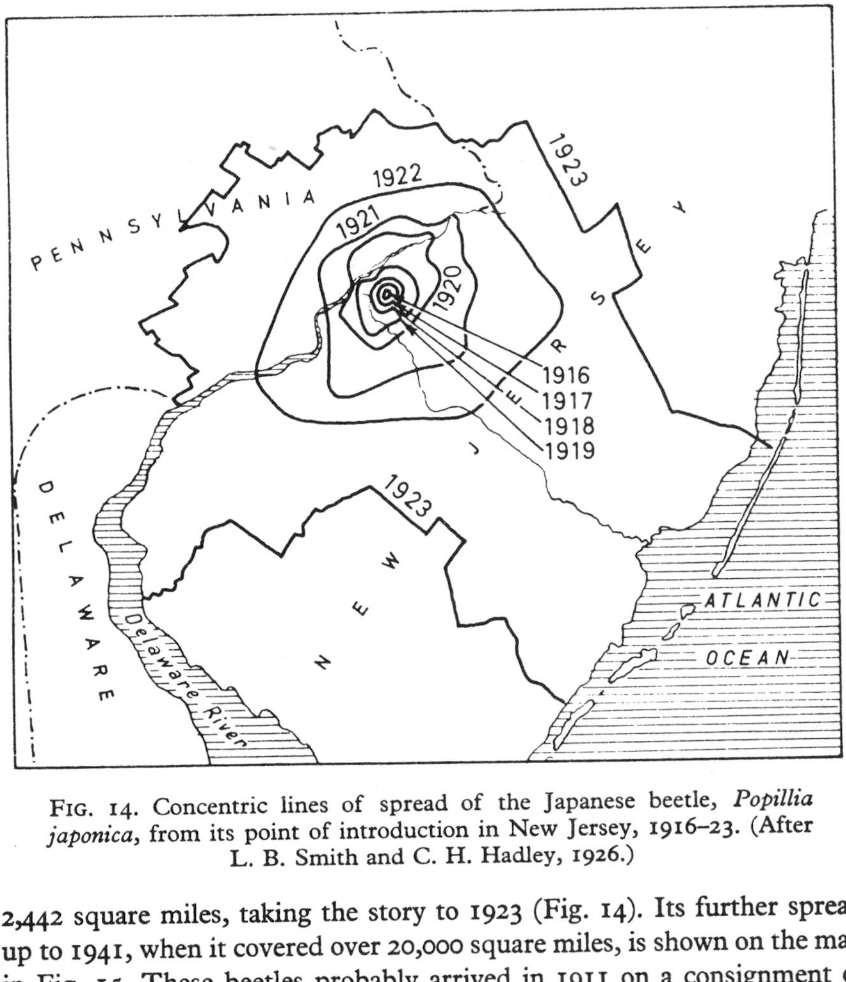

FIG. 14. Concentric lines of spread of the Japanese beetle, *Popillia japonica*, from its point of introduction in New Jersey, 1916–23. (After L. B. Smith and C. H. Hadley, 1926.)

2,442 square miles, taking the story to 1923 (Fig. 14). Its further spread up to 1941, when it covered over 20,000 square miles, is shown on the map in Fig. 15. These beetles probably arrived in 1911 on a consignment of iris or azaleas from Japan. In Japan they are seldom a pest, but in America the numbers at first were formidable.[66] By 1919 a single person could gather up 20,000 beetles in a day; in one orchard containing 156 not very old peach trees, 208 gallons volume of the beetles was taken in two hours, and next day it was said that the numbers on the trees appeared unchanged! A beetle population that will feed on and often defoliate over 250 species of

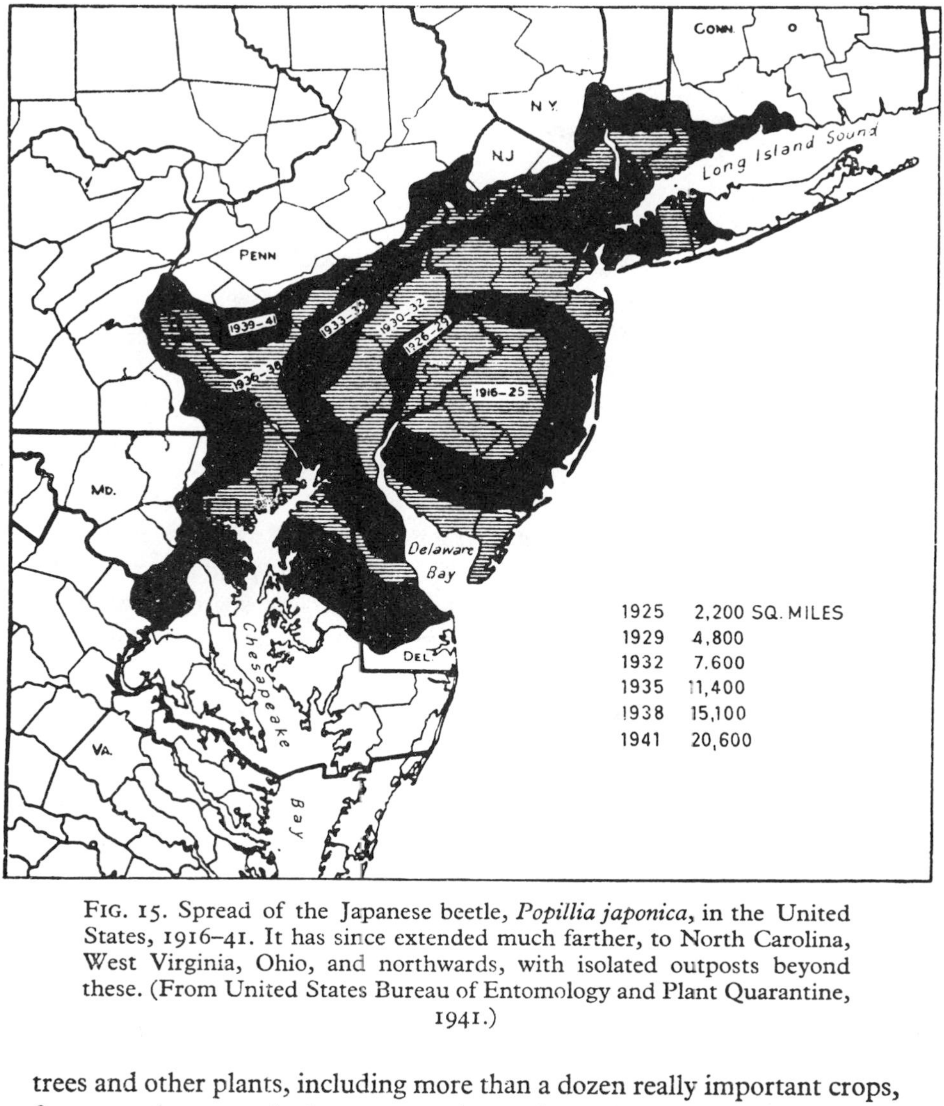

FIG. 15. Spread of the Japanese beetle, *Popillia japonica*, in the United States, 1916–41. It has since extended much farther, to North Carolina, West Virginia, Ohio, and northwards, with isolated outposts beyond these. (From United States Bureau of Entomology and Plant Quarantine, 1941.)

trees and other plants, including more than a dozen really important crops, from soy beans and clover to apples and peaches and shade trees, is portentous.[64] It is matched by another Oriental insect that came on nursery stock from Japan in the early nineteen-twenties: the camphor scale insect, *Pseudaonidia duplex*. This has invaded Louisiana, Texas, and Alabama. There are nearly 200 host plants upon which it can live, though citrus trees are the ones that matter most to agriculture.[72] But in Japan, the camphor scale is not a serious pest; the same can be said of the Asiatic garden beetle, *Autoserica castanea*, which lives in China and Japan, but reached New Jersey in 1921, and has since spread as a garden pest around New York City.[86] Larvae of this chafer, the Japanese beetle, and also the Oriental beetle, *Anomala orientalis*, another new invader in the nineteen-twenties, may all be found living together in turf in that area.[58]

Introductions come from all parts of the world. Perhaps nearly half the 180 or so major plant pests of the United States are from abroad.[100] By far the greater number of invasions in North America have been from Europe, as is to be expected from the heavy traffic over such a long period. Two of the earliest were the Hessian fly, *Mayetiola destructor*, on wheat and the codling moth, *Carpocapsa pomonella*, mainly in apple orchards, and after them a further long procession of immigrants.[90] Some, like the clover root borer weevil, *Hylastinus obscurus*, which arrived about 1878, more or less covered the country (as with the starling, and the house sparrow before it).[98] Others, like the European corn borer moth, *Pyrausta nubilalis*, which started in 1917, have moved fairly slowly and steadily but are probably still expanding.[59,106] Some comparatively recent introductions may become much more important in the future: among these are the European cockchafer beetle, *Melolontha vulgaris* or—the name used in America—*Amphimallon majalis*, which reached New York State in 1942;[85] and the golden nematode, *Heterodera rostochiensis*, that was noticed about the same time on Long Island but probably arrived ten years before that.[65] These two species damage crops both in Europe and America, the former having larvae that live at roots and adults that defoliate trees, and the latter damaging potatoes. At present the nematode occupies only about 8,000 acres of potato land, yet cannot so far be eradicated.

Turning now to the other continents that have sent their contingents to North America, we may notice the vegetable weevil, *Listroderes obliquus*, and the Argentine ant, *Iridomyrmex humilis*, both from South America. The vegetable weevil reached the United States in 1922 and has occupied three of the Southern states;[87] it has also worked its way to Australia and

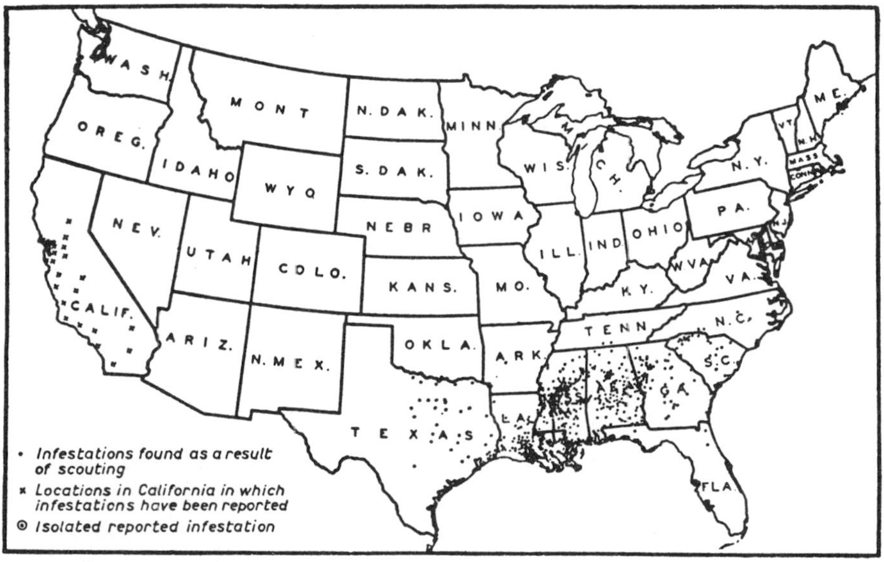

FIG. 16. Distribution of the Argentine ant, *Iridomyrmex humilis*, in the United States. (From M. R. Smith, 1936.)

South Africa. The ant was first noticed at New Orleans, Louisiana, in 1891, but must have got there some time before that, possibly on ships bringing coffee from Brazil. This extraordinarily aggressive ant has its natural home in South America and was first described from the Argentine and later from Brazil, Uruguay, and elsewhere. Newell and Barber remarked: 'That Argentina is its native home is also borne out by the fact that it does not appear to be generally a pest of importance in that country.'[94] In less than fifty years from its introduction at New Orleans the ant had invaded a large part of the southern States, and also arrived (by 1905) in California, where it became widely distributed (Fig. 16). Everywhere it multiplied immensely and invaded houses and gardens and orchards, eating food or—out-of-doors—other insects, also farming scale insects and aphids on various trees to and from which the ants march along trackways, just as do our English wood ants to trees like pine and birch. They also go into beehives to take the honey. A conspicuous character of this fierce and numerous tropical ant is that it drives out native ants entirely. Smith 'often witnessed combats in the field between native and Argentine ants . . . The fact that the Argentine ant destroys practically all the native ants as it advances makes it comparatively easy to delimit an area infested by them . . . Just as soon as the Argentine ants begin to disappear, native ants invade the territory, and within a few years are as plentiful as ever.'[103] So might the wolves and foxes and jaguars have advanced into South America in Pliocene times, driving out the native borhyaenid marsupial carnivores. The Argentine ant is not, as a matter of fact, a very fast natural invader, for its nuptials take place almost entirely within the nest, and its movements by crawling would not take it more than a few hundred feet a year. It seems that transport in merchandise, especially by railway train, dispersed it so quickly within the United States.

The Argentine ant has also spread to other countries in an explosive way. In South Africa and Australia there has been the same elimination of native ants. Australia both in the south and west was reached by 1939–41, and a further bridgehead in New South Wales by 1951. In 1955–6 the areas covered by the ant were about forty-two square miles in Western Australia, ten in Victoria, and three and a half in New South Wales. Though poison baits had somewhat mitigated the American invasion, in Western Australia the ants did not accept baits, but the use of contact insecticides like DDT and dieldrin has already been very successful. Houses sprayed with strong concentrations of dieldrin remained lethal to ants for as long as four and a half years and the campaign to wipe out the ants altogether is still in full swing. In the places that have been cleared (and, incidentally, the spraying kills not only ants but all flies, mosquitoes, cockroaches, and fleas—and what else?), the ousted species of native ants quickly reappear and occupy it, as soon as the poison has gone.[84]

The supreme example of a species introduced from Australia is the fluted or cottony cushion scale insect, *Icerya purchasi*, which appeared in California about 1868 and thereafter began to threaten the whole future of its citrus orchards. It is a famous insect among economic entomologists because it was controlled completely in a couple of years by the very numerous descendants of 139 specimens of the ladybird *Novius* (*Vedalia*) *cardinalis*, a native enemy of the fluted scale in Australia. Australia administered the poison, but it also supplied the antidote, and this miracle of

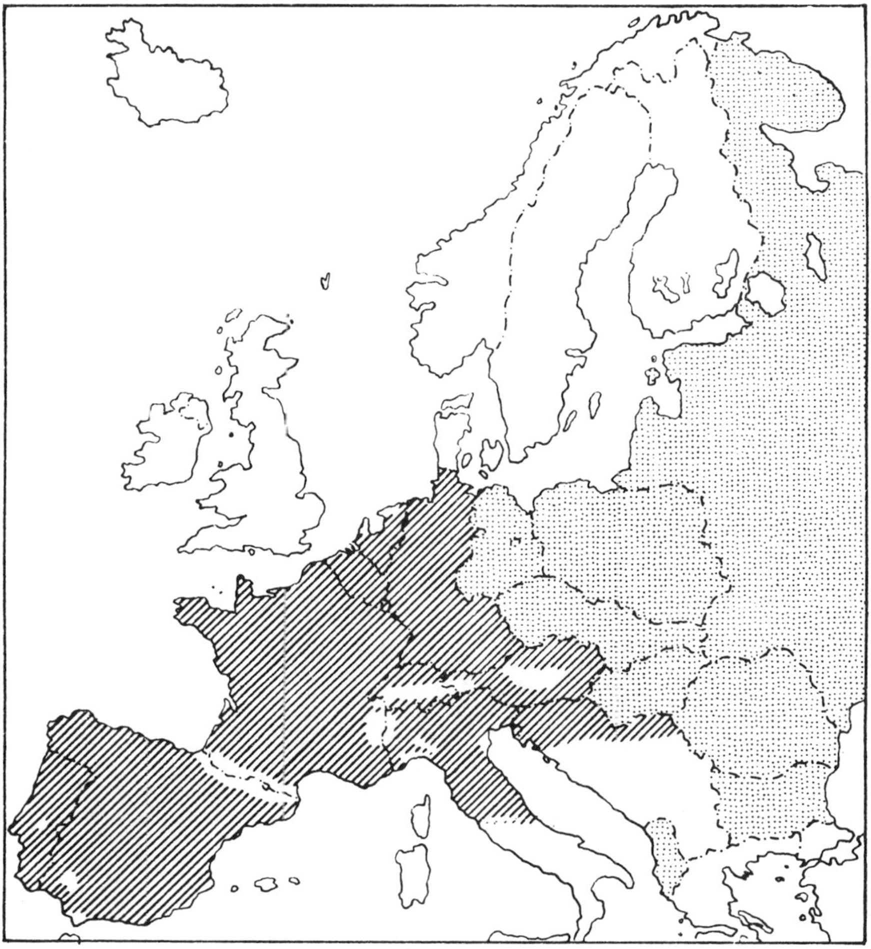

FIG. 17. Distribution of the Colorado beetle, *Leptinotarsa decemlineata*, in Europe in 1956. *Cross hatching:* present occupied areas; *white:* free (except for a few isclated outbreaks); *Stipple:* no information. (From information supplied by the European and Mediterranean Plant Protection Organization.)

ecological healing was afterwards performed in every other country to which the scale insect came and began to be a pest—as Europe, Syria, Egypt, South Africa, Japan, Hawaii, New Zealand, and South America. Of this discovery—the idea came from Riley and the field work was done by Koebele—Howard remarks: 'So striking a success may probably never again be achieved in this country', and adds that it raised too much optimism about the ease with which the introduced enemies and parasites could be used to combat invasions from abroad.[90] For example, the careful and intelligently planned introduction of parasites and predators from Japan to control the Japanese beetle has not acted in the wholesale and concentrated manner of *Novius cardinalis* on the fluted scale.

The transatlantic movement of the herbivores of crops has not only

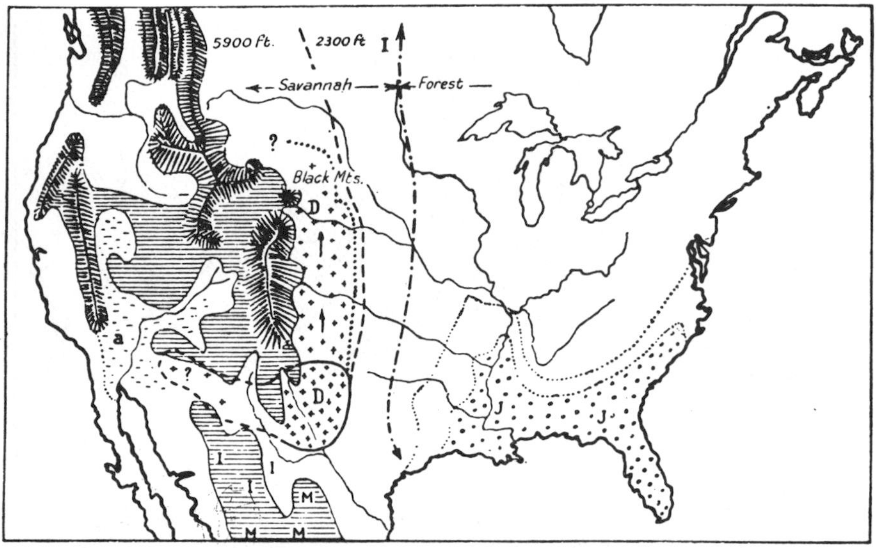

FIG. 18. Natural distribution of the Colorado potato beetle, *Leptinotarsa decemlineata*, in Western North America, before it colonized the cultivated potato. Originally found in New Mexico and Arizona, it moved slowly northwards at the end of the eighteenth century into the prairies along the east of the Rocky Mountains (crosses and D). Limits to the west were the arid valley of the Colorado River, and the Rocky Mountains; to the east an area with no *Solanum* species. I, J, and M are three other species of *Leptinotarsa*. (From B. Trouvelot, 1936.)

been one way. About ninety years ago France was invaded by the American vine aphid, *Phylloxera vitifolii*, which had a quiet home on wild vines in the United States east of the Rocky Mountains. After its entrance by Bordeaux and perhaps by some other ports as well, it very soon was spread into the wine-growing parts of Europe and also to Algeria, brought on vine-stocks and often spread by these from place to place.[90] This *Phylloxera* has also occurred at least eighteen times in the British Isles.[110] On European vines its root galls were fatal. After the nadir of the wine industry, with three million acres of French vineyards destroyed, a Frenchman had the idea of grafting the European vines onto American rootstocks resistant to the root-phase of the *Phylloxera*. With this discovery, the economic danger passed. Australia had it in 1875 and California also by the eighteen-nineties.

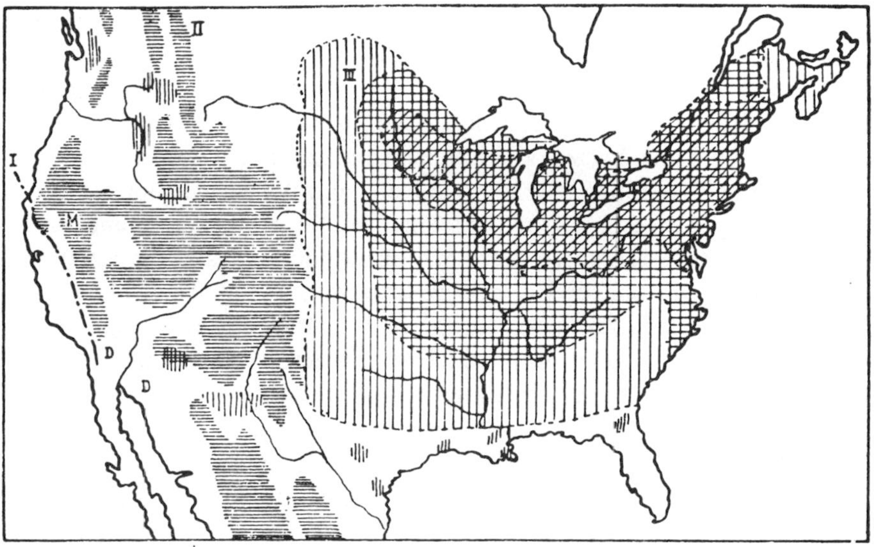

FIG. 19. Invasion of the Colorado beetle, *Leptinotarsa decemlineata*, onto cultivated potatoes in North America (*vertical hatching*). The northern area (*cross and slanting hatching*) has serious damage, the zone round this (*cross hatching*) important but not intense damage. The spread westwards has partly been hindered by high mountains (*horizontal hatching*) and desert (D), the dot-dash line giving the extreme limit. (From B. Trouvelot, 1936.)

Another famous invader also came from the United States—the Colorado beetle, *Leptinotarsa decemlineata*, which lives naturally in the eastern part of the Rocky Mountain region from Colorado south to Mexico, feeding chiefly upon the wild sand-bur, *Solanum rostratum* (Fig. 18). It will also eat other species of the potato family, and a few plants of different sorts as well.[104] When the beetle itself was discovered by entomologists the cultivated potato, *S. tuberosum*, had not yet been brought to Colorado, and the beetle only began to spread when its new future habitat—the potato crop—had spread westwards to touch its natural distribution. After this, from 1859 for about twenty years the beetle population spread eastwards, and by 1874 the Atlantic shore was reached (Fig. 19). For a time it was thought that potatoes could not be grown in the region of its advance. Its first bridgehead across the ocean was in Germany, in 1876, but it was destroyed. Later small invasions came to nothing, including one at Tilbury in 1901—until 1920, when the population arriving (like the *Phylloxera*) at Bordeaux from abroad overcame control and by 1935 almost the whole of France was occupied,[109] with subsequent spread to other countries (Fig. 17). Intermittent advances across the English Channel (264 outbreaks in 1947) so far have been subdued.[88] In 1955 England and Wales had about 605,000 acres under potato crops. Only fifty-three beetles were found, and these were intercepted, for there were no inland outbreaks. Scotland with 154,000 acres had one beetle, and Ireland with 117,000 had none.[75] A Nearctic *Solanum* is taken to Europe in the sixteenth and seventeenth centuries, bred into new forms which return across North America and start the beetle population moving eastward, eventually to occupy Europe. But the herbivore is still out of balance with its food plant.

The latest big invader from North America is the moth known as the fall webworm, *Hyphantria cunea*, which reached Hungary during the Second World War, in 1940. After a few years it spread fast into Austria, Czechoslovakia, Roumania, Yugoslavia, and parts of the Ukraine.[76] This caterpillar can completely defoliate many kinds of trees and plants, and is now doing so in Europe. It does not invade the forests but stops at the edge of them—at present. One of the peculiar things about this invasion is that in Europe the caterpillars have a strong preference for mulberry trees, which they hardly touch in America.

A disadvantage of describing invasion only by examples, however famous, is that this does not quite convey the tumult and pressure of species that have been and are escaping from the confinement of their ancestral continents to range the world. We might really use the words of Walt Whitman in his poem suitably entitled 'As consequent, etc.':

Some threading Ohio's farm-fields or the woods,
Some down Colorado's cañons from sources of perpetual snow,
Some half-hid in Oregon, or away southward in Texas,
Some in the north finding their way to Erie, Niagara, Ottawa,
Some to Atlantica's bays, and so to the great salt brine.

Whenever we know the history it starts with a very small nucleus of population, growing to an 'Autumn Rivulet' and then not infrequently to a flood. And when the population has got that far, its movement is seldom absolutely checked except by natural limits of the environment. The historical movements of crop pest invasions in the world are illustrated by six maps (Figs. 20–5) from the fine series compiled by the Commonwealth Institute of Entomology.[68] (The Commonwealth Institute of Mycology publishes a similar series for the fungus diseases of crops and trees.) These maps show several kinds of stages in the spread of species. Of course many are still confined to their original continental home, even if they have expanded widely within it, as agriculture itself expanded. The Japanese beetle (Fig. 20), and the European spruce sawfly, *Gilpinia hercyniae* (Fig. 21), are Palaearctic species spread to the eastern part of North America. The former is East Asian, the latter European. The Colorado beetle, as has been described, illustrates the reverse movement (Fig. 22). The lucerne flea, *Sminthurus viridis*—not a flea but a springtail—has reached the Antipodes but not North America (Fig. 23). The small cabbage white butterfly, *Pieris rapae*, is also a Palaearctic species, which spread in the middle nineteenth century across North America, and has also solidly established itself in Bermuda, Australia, New Zealand, and Hawaii (Fig. 24). The Australian fluted scale insect seems to have travelled to almost every country that it can occupy (Fig. 25). One sees that, although the eventual result may be the same in other species, the process has still far to run, particularly since Wallace's Realms are now stoutly defended

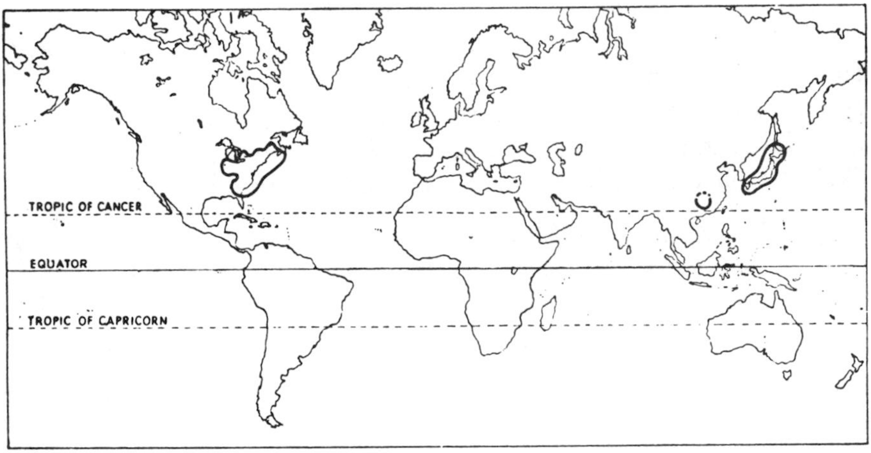

FIG. 20. The Japanese beetle, *Popillia japonica* (up to 1952).

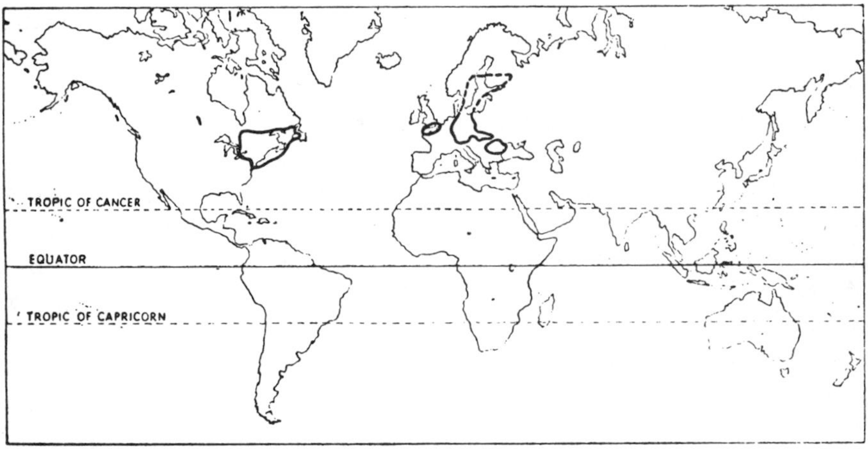

FIG. 21. The European spruce sawfly, *Gilpinia hercyniae* (up to 1953).

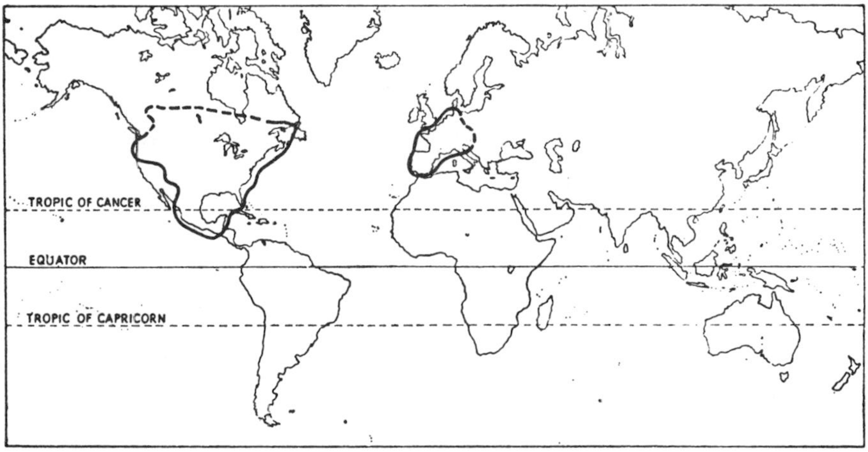

FIG. 22. The Colorado potato beetle, *Leptinotarsa decemlineata* (up to 1951).

FIGS. 20–2. Different stages and directions in the break-down of Wallace's Realms by insect pests. (By courtesy of the Commonwealth Institute of Entomology.)

by massive quarantine systems and plans for eradication. Yet, in spite of quarantines, at any rate in the United States, a very serious pest of cotton crops was able to get to four new continents within about twenty-five years. This is the pink bollworm, *Pectinophora gossypiella*, a small brown moth whose later larvae are coloured pink, that probably lived originally in India and perhaps South-east Asia generally, on wild and cultivated cotton. It may have been in Central Africa before its world spread started. But it is thought to have been first brought to Egypt on imported cotton or cottonseed from India in 1906, attracted notice there in 1911, and spread to East and West Africa; also to China, to various islands like the Philippines and Hawaii; from Hawaii to the West Indies; to Mexico in 1911, and Texas (on Mexican cottonseed) by 1917. Later on it reached Brazil and Australia.[91]

It will be noticed that invasions most often come to cultivated land, or to land much modified by human practice. Yet there are some other species—still a minority—that penetrate further, into natural waters and woodlands, into communities that are at any rate rich and varied even if they have also suffered the results of human occupation through fire, forest

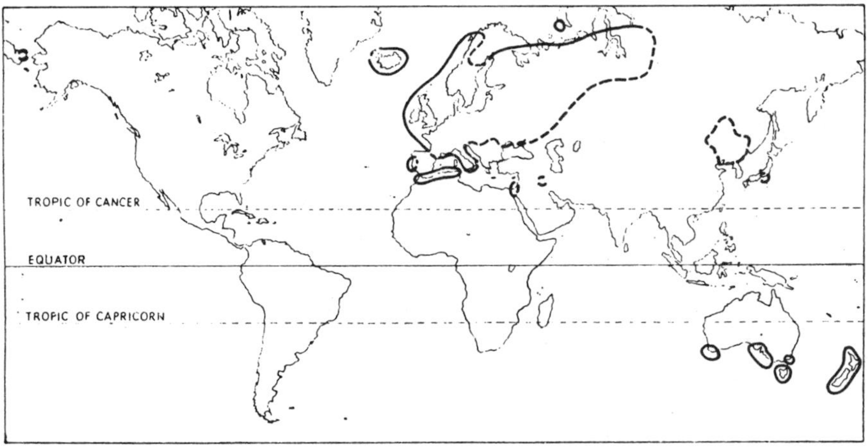

FIG. 23. The lucerne 'flea', *Sminthurus viridis*, from Europe (up to 1956).

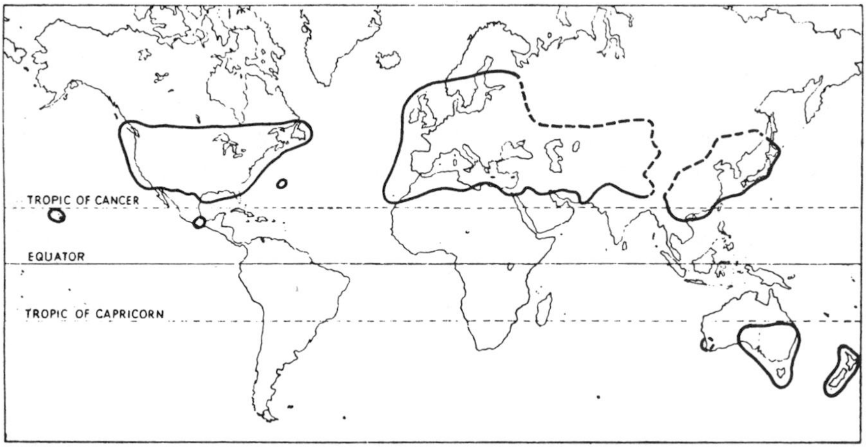

FIG. 24. The small cabbage white butterfly, *Pieris rapae*, from Europe (up to 1952).

succession after lumbering, water control or channel drainage. Amongst these species I have already described the North American muskrat, the South American coypu, and the Chinese mitten crab, and mentioned the American grey squirrel. We may supplement these with examples from a small but powerful contingent of species brought from Europe and accidentally introduced into the forest lands of North America.[71] There are, first of all, those like the Japanese beetle already described, that attack garden and shade trees rather than the natural forests. One group of moths that have arrived in succession and invaded the eastern forests includes the gypsy moth in 1869; the brown-tail moth, *Nygmia phaeorrhoea*, in 1897; and quite recently, about 1949, the winter moth, *Operophtera brumata*. Other broad-leaved trees have also acquired new invaders: two kinds of leaf-mining sawflies on birch, *Phyllotoma nemorata* and *Fenusa pusilla*, the former first noticed in 1905, the latter not well dated; the satin moth, *Stilpnotia salicis*, on poplars and willows since 1920; the small green willow beetle *Plagiodera versicolora* (common enough skeletonizing leaves of the pollard willows along English rivers, and in America[89] doing this to both

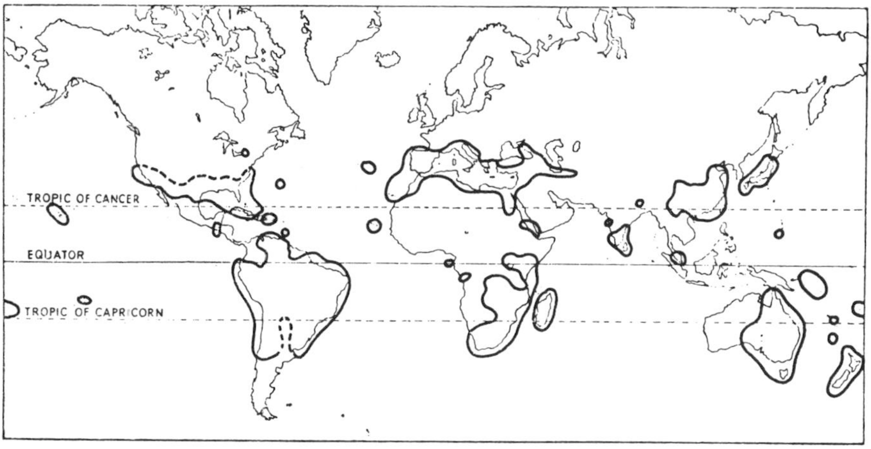

FIG. 25. The fluted scale insect, *Icerya purchasi*, from Australia (up to 1955).

FIGS. 23–5. Different stages and directions of the break-down of Wallace's Realms by insect pests. (By courtesy of the Commonwealth Institute of Entomology.)

native and imported species of willows) in 1911 onwards; the elm bark-beetle, *Scolytus multistriatus*, in 1909 and the elm leaf beetle, *Galerucella xanthomeleana*, about 1840—though the leaf-beetle does not go right into forests; and the felted beech scale insect, *Cryptococcus fagi*, by 1890.[81] On coniferous trees there are also tremendous invasions still in progress: the European spruce sawfly, *Gilpinia*, or *Diprion*, *hercyniae*, since at least 1922; three kinds of pine sawflies, *Diprion simile* in 1914, *Neodiprion sertifer* in 1925, and *D. frutetorum* not well dated; the balsam woolly aphid or fir bark louse, *Adelges* or *Chermes piceae*; the larch case-bearer moth, *Coleophora laricella*, since 1909; the European pine shoot moth, *Rhyacionia buoliana*, in the United States by 1914, and Canada 1925.[69]

It used to be taken for granted that the larch sawfly, *Pristophora erichsonii*, which is a tremendous forest pest in North America, also came from Europe. There is no doubt that it now has a Holarctic range, as does the larch, of which there are four species, in North America, Siberia, Japan, and Europe. But Coppel and Leius conclude that the evidence 'indicates only that the larch sawfly has been in North America for some time. Its origin cannot be determined on the basis of evidence at hand.'[70] Perhaps it arrived in very early days, like the Hessian fly.

Some excellent records and maps of invasion have been made by forest entomologists in Canada and the United States, of which four are selected here (a fifth, of the gypsy moth, comes in Chapter 6). The winter moth is such a new arrival that it has hardly had time to get into the text-books.[79,82] It has little more than a bridgehead in Nova Scotia (Fig. 26). The felted scale insect of beech has centres at various places. Its slow spread on a straight front in Nova Scotia (Fig. 27) from the original entry at Halifax before 1890 has been remarkably regular.[81] This insect helps the natural inoculation of the woody tissues of beech with a fatal fungus, *Nectria*, that causes cankers in the trees. In Nova Scotia in 1948 over 80 per cent. of the beech trees were cankered, and things were as severe on Prince Edward Island. This felted scale can commonly be seen in the cracks of bark on some British beeches, but only in Denmark and eastern America do the ecological conditions result in fatal disease.

The elm disease also comes from the combined action of insects and a species of fungus, *Cerastomella ulmi*. It has injured and killed many trees

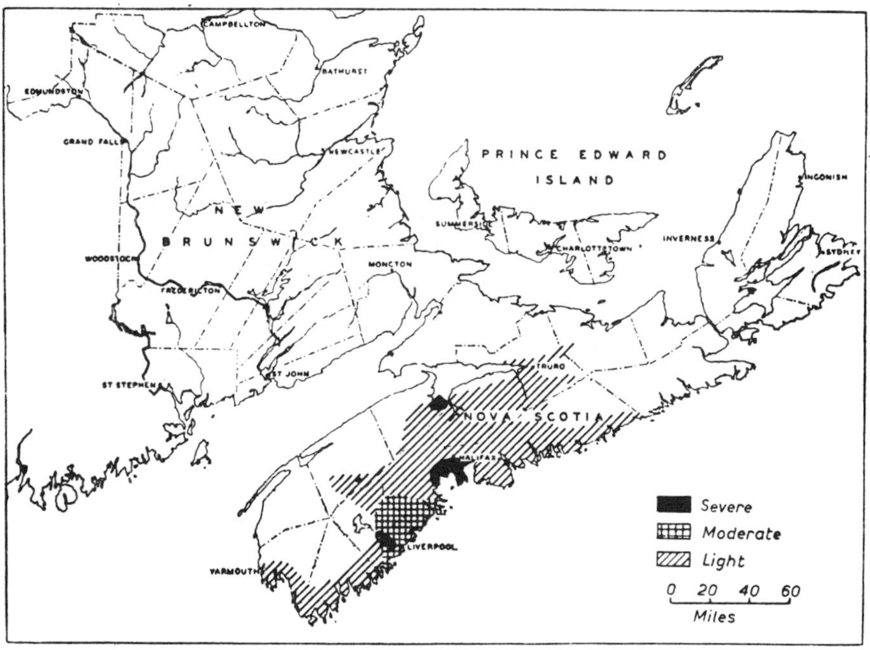

FIG. 26. The start of a new continental invasion: the winter moth, *Operophtera brumata*, arrived from Europe about 1949 and is spreading in Canada. (From the Forest Insect and Disease Survey, Dept. of Agriculture, Canada, 1956.)

in Europe: its original home is not known. Since 1927 it has also spread across England, under the name of 'Dutch elm disease', at first killing an alarming number of hedgerow elms, but now in a chronic state from which only local epidemics flare up from year to year in different parts of the country.[95] These dead elms can often be seen (Pl. 20). In England the fungus is spread and partly inoculated into the tissues of the trees by two species of bark-beetle, *Scolytus multistriatus* and *S. destructor* (= *scolytus*): this happens because the adult beetles feed on the fresh bark of twigs, often in a fork or crutch, and will do this on quite healthy trees, though their breeding galleries under bark are usually in less healthy, dying, or felled trees.[83] One of these bark-beetles, *multistriatus*, though not the other, reached the United States early in this century, probably in unbarked elm

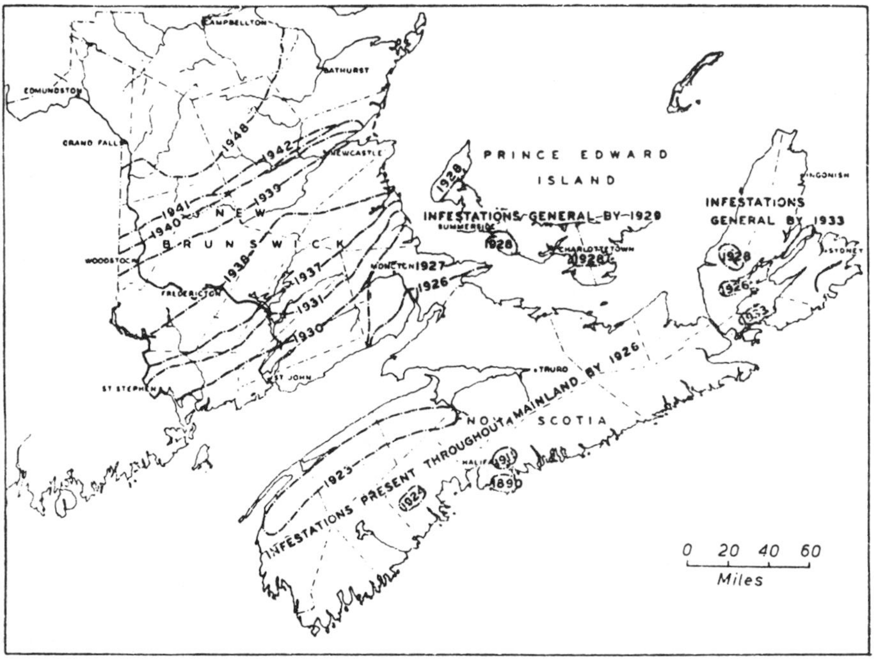

FIG. 27. A slow continental invader from Europe to Canada: the felted beech scale insect, *Cryptococcus fagi*, after about sixty years. (From the Forest Insect Survey, Dept. of Agriculture, Canada, 1949.)

timber brought to seaports and perhaps also carried inland.[67] The first record was in Massachusetts in 1909, and the beetle has since then spread to many parts of New England, its distribution in 1938 looking as if it had started from two main centres, one in the New York region and the other, almost joined to it, from southern New Hampshire. Throughout and beyond this range there lives a native elm bark-beetle, *Hylurgopinus rufipes*, that can also carry the fungus, though it may not be such an effective agent as the European one. Recently, a third bark beetle, *Scolytus sulcatus*, that had not been recorded for many years, was found to be quite widespread.[96] The last is a native species that chiefly lives on apple trees, but also comes on elm, though its role in the disease, if any, is not much known so far. The disease itself, that is the combination of elm, bark-

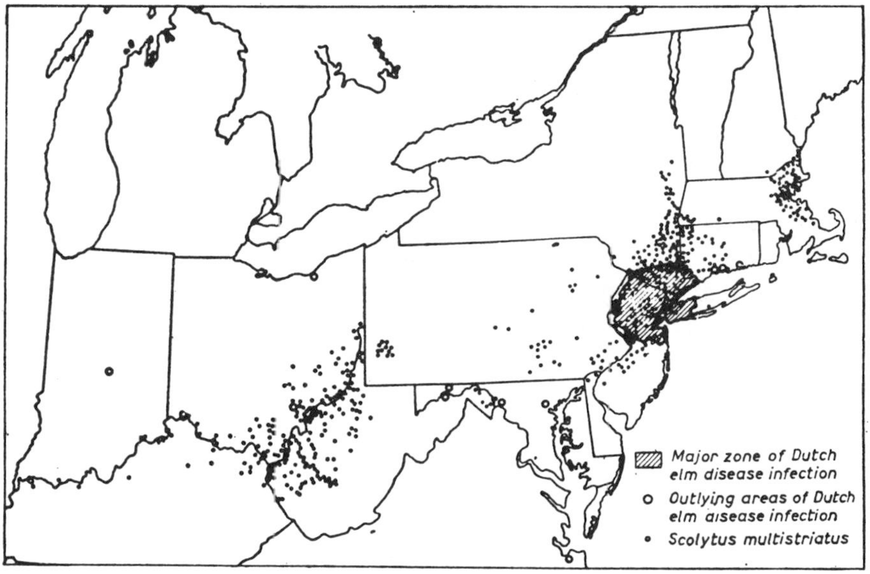

FIG. 28. The invasion of a disease-carrying species, the elm bark-beetle, *Scolytus multistriatus*, from Europe to North America, and the later and more limited spread of the fungus, *Cerastomella ulmi*, that causes elm disease. Distribution by 1937. (From C. W. Collins, 1938.)

beetles, and fungus, has a much more restricted distribution than that of the invading beetle (Fig. 28). The fungus is thought to have entered with infected elm timber used for veneers, and it was first identified in Ohio in 1930. Between 1934 and 1940 the main invasion area of the disease increased from about 2,500 square miles to nearly 11,000, and beyond this were scattered points as well. This main area in 1940 comprised half of New Jersey, the south-east corner of New York and part of eastern Connecticut. Ecologically, this big invasion is interesting because the insect vector seems to have arrived and spread in advance of the fungus; just as the mosquito carrier of yellow fever, *Aedes aegypti*, has a range far beyond the present occurrence of this virus, which has not yet reached Asia. The destruction of elms by the disease, as well as by the measures used for control, has been enormously extensive. Brewer reported that from 1933 to 1940 four and a quarter million elms had been removed in the course of tree sanitation, and yet that 'it cannot be stated that the Dutch elm disease is being eradicated in the major region', though some outlying points of invasion had been mastered.[62] Lest these figures give a wrong impression of utter destruction of the species of elm there, it should be added that the average number of the remaining trees found with infection in this major region was only 1 in 8,000—but this also conveys a good idea of the importance of that tree for shade and forest in America.

We come now to the invaders of conifer forest. In 1941 Brown published a list of 101 species of foliage-eating insects living on spruce in Canada.[63] Nearly all these are native forms, a few of which, like the spruce

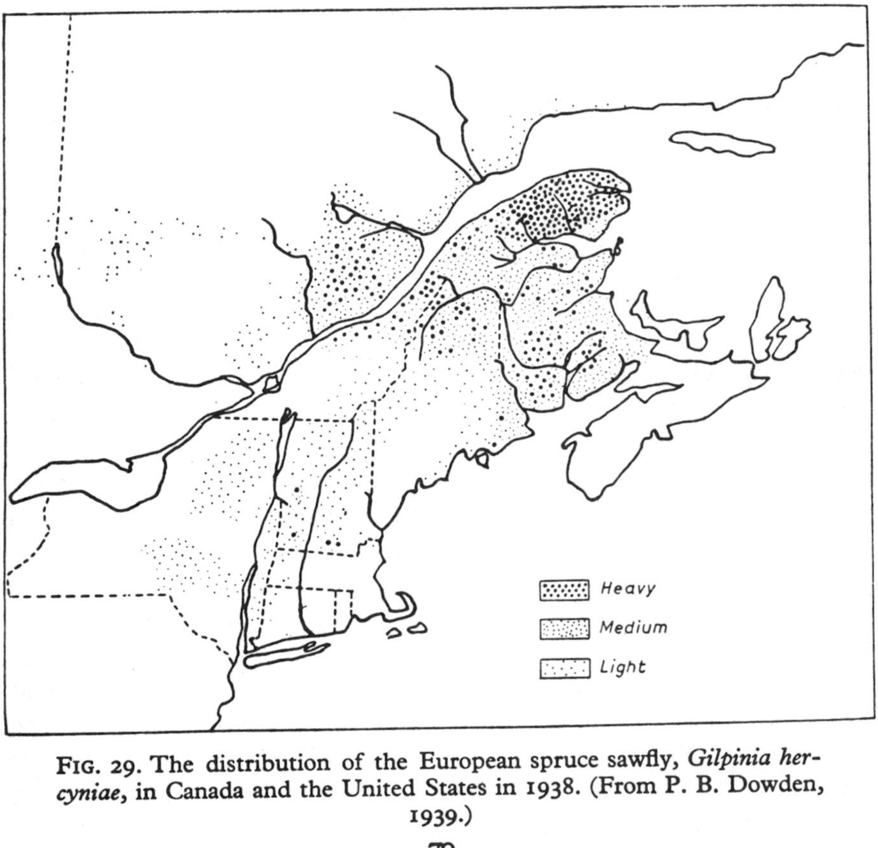

FIG. 29. The distribution of the European spruce sawfly, *Gilpinia hercyniae*, in Canada and the United States in 1938. (From P. B. Dowden, 1939.)

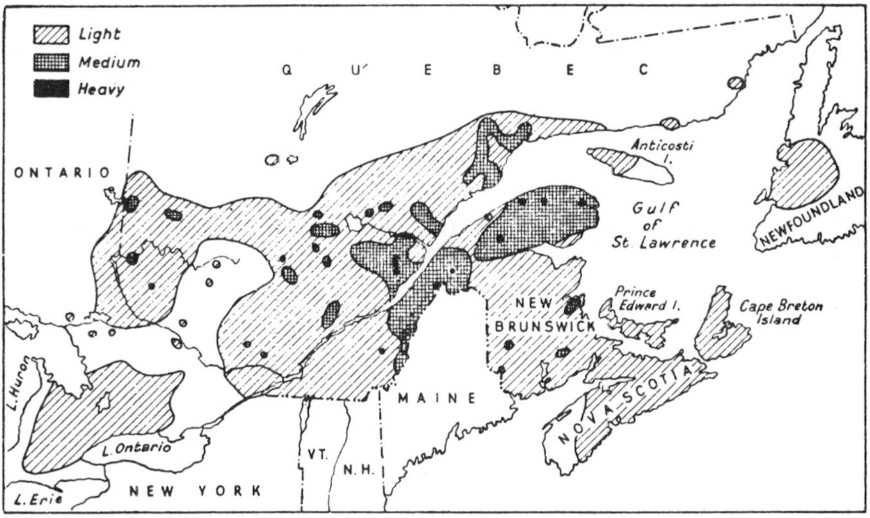

FIG. 30. A detailed map of the distribution and intensity of European spruce sawfly populations in Canada in 1942. The intensity shows a strong decline from the catastrophic numbers a few years before. (After the Forest Insect Survey, Dept. of Agriculture, Canada, 1943.)

budworm moth, *Archips fumiferana*, do serious harm. In 1930 a European species of spruce sawfly was found defoliating the white and black spruce in the Gaspé Peninsula of southern Quebec. It is known to have arrived in North America in the twenties and may have been present earlier than that.[74] By the time the Gaspé eruption was studied, nearly two-thirds of the white spruce (Pl. 21) and a quarter of the black spruce had been killed on an area of 3,000 square miles, and infestations were also spread beyond this into other parts of Quebec, the Maritime Provinces of Canada, and New England down to New York. By the end of 1937 the Gaspé Peninsula had heavy populations on nearly 10,000 square miles, and Balch remarks that by then this sawfly was the most abundant spruce-feeding insect in north-eastern America (Figs. 29, 30). Because of discrepancies between the habits in Europe and America a careful examination was made[60] which revealed that there are two species in Europe, *polytomum* and *hercyniae*, and that it is the latter that has been brought into America; but earlier reports often use the name *polytomum*. Spruce forms more than a fifth of the timber of Canada, and this sawfly feeds on most of the species, though on nothing but spruce. Here is a foreign invader taking a dominant position in the community of natural forest—though the 'natural' forest of Canada has been so highly modified by a long history of lumbering and fire that the word needs to be used mainly in the sense that any forest system has a more complex ecology than any field crop, and therefore many of the characteristics of virgin vegetation. According to Dowden: 'A striking feature of the spruce sawfly outbreak in Canada and the United States has been the almost total absence of attack by parasites, although a number of predators, such as shrews, mice and squirrels, may destroy up to 50 per cent. of the hibernating cocoons. In Europe on the other hand, where the insect has been known for over 100 years and has caused little or no damage, it has a number of valuable parasitic enemies.' In 1932 the introduction of its European parasites began, with the release in due course of more than twenty species. The scale of these operations was gigantic: about seven million *Microplectron fuscipennis*, a gregarious chalcid parasite of the cocoons, were liberated! Since then hundreds of millions more have also been bred for this purpose, though the final outcome of the operation has still to be assessed.[61a]

The North American forests, like their orchards and gardens, are also beginning to receive the first trickle of invaders from Asia. Here I will only mention the bark-beetle *Xylosandrus germanus* from Japan and China, which arrived in the eastern United States by 1932 and in Germany twenty years after that. It burrows in trees like alder, beech, and oak and its invasions have really only started.[107,108]

Stored grain and other warehouse and manufactured stuff are accumulating new inhabitants from other continents, though many of the alien insects and mites have spread so long ago, when records were not kept or the species not correctly classified, that their place of origin is not very easy to pin down. A large number are now almost cosmopolitan, and because they inhabit buildings include subtropical species able for this reason to survive in temperate latitudes; just as our hothouses contain many species that live naturally in the tropics. But among the inhabitants of stored products in Britain that are still newcomers are the Australian

14. A forest scene on the upper Amazon. From left to right: umbrella bird, *Cephalopterus*; two perched curassows, *Crax*; two flying curl-crested toucans, *Pteroglossus*; a trumpeter, *Psophia*; two whiskered humming-birds, *Lophornis*. Behind, a jaguar. (Drawing by J. B. Zwecker, in A. R. Wallace, *The Geographical Distribution of Animals*, 1876.)

15. Mammals of the North American prairie. Behind, a herd of bison and two prong-horned antelope, *Antilocapra*; on the left, a pocket gopher, *Geomys*—which in fact lives almost entirely hidden underground; and on the right three prairie dogs, *Cynomys*. (Drawing by J. B. Zwecker, in A. R. Wallace, *The Geographical Distribution of Animals*, 1876.)

16. Mammals in Western Tartary (Central Asia). Behind, wolves and three saiga antelope; on the left a subterranean mole-rat, *Spalax*; and on the right, a desman, *Myogale*, a large water insectivore over a foot long. (Drawing by J. B. Zwecker, in A. R. Wallace, *The Geographical Distribution of Animals*, 1876.)

17. A river and forest scene in equatorial West Africa. From left to right: a water insectivore, *Potamogale*; two gorillas; on the branch a plantain-eater, *Turacus*; flying, a whydah finch, *Vidua*; couched, a red river-hog, *Potamochoerus*; above in the tree, a potto—a lemur—*Perodicticus*. (Drawing by J. B. Zwecker, in A. R. Wallace, *The Geographical Distribution of Animals*, 1876.)

18. A scene on the plains of New South Wales. From left to right: two kangaroos, *Macropus*; a pair of lyre-birds, *Menura*; two emus, *Dromaeus*; two crested pigeons, *Ocyphaps*; a frogmouthed goat-sucker, *Podargus*; and a gliding opossum (a phalanger), *Petaurus*. (Drawing by J. B. Zwecker, in A. R. Wallace, *The Geographical Distribution of Animals*, 1876.)

19. Birds in a Malay Peninsula forest. From top to bottom: three white-handed gibbons, *Hylobates*; left, two broadbills, *Corydon*; right, a drongo shrike, *Edolius*; an argus pheasant, *Argusianus*, displaying to a hen; and a rhinoceros hornbill, *Buceros*. (Drawing by J. B. Zwecker, in A. R. Wallace, *The Geographical Distribution of Animals*, 1876.)

20. A dying English elm in Oxfordshire, with undamaged ones close by. To the left of the dying tree is the bole of another of which the dead top has broken off in the past. The disease is caused by a fungus, *Cerastomella ulmi*, spread by bark-beetles. (Photo C. S. Elton, 1957.)

21. Destruction of white spruce, *Picea glauca*, by the European spruce sawfly, *Gilpinia hercyniae*, near the head of the Cascapedia River, Quebec, October 1932. (The dead and dying spruce show grey; the dark trees are healthy balsam firs, *Abies balsamea*.) (By courtesy of the Science Service, Canada Dept. of Agriculture; details from R. F. Morris, Forest Biology Laboratory.)

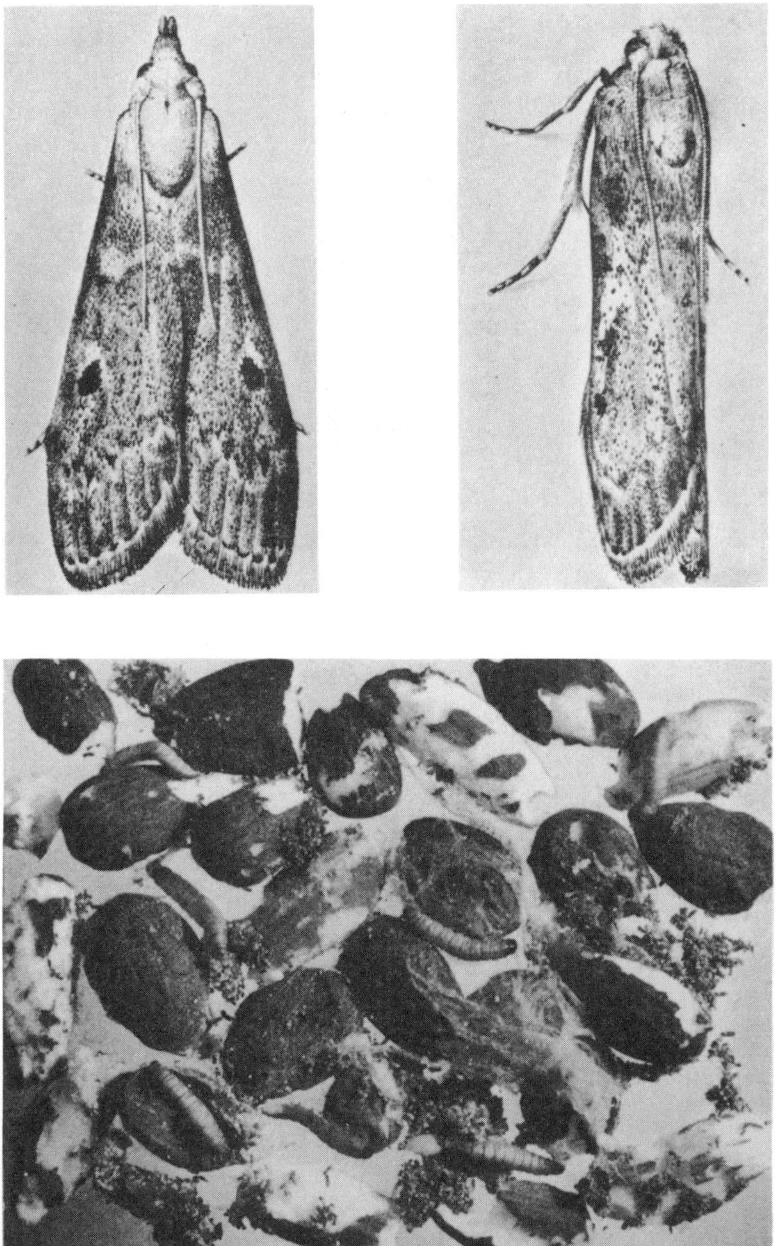

22. A galleriid moth, *Aphomia gularis*, that has spread in recent years from the Orient to Europe and North America in stored products. Above, the adults; below, the caterpillars eating stored almonds. (From K. G. Smith, 1956, by permission of the Controller, H.M. Stationery Office. Crown Copyright.)

carpet beetle, *Anthrenocerus australis*, which has been here since 1933—though chiefly in seaports—and may become one of our textile and household pests as it is in Australia and New Zealand;[61] and the moth *Aphomia gularis* from the Oriental Region, now spread to Europe and North America though also so far chiefly in the coastal parts (Fig. 31), where its larvae devour such things as stored almonds, walnuts, groundnuts, and prunes (Pl. 22). It is not yet known to have reached the southern continents.[101]

Nearly all the insect immigrants I have been discussing were introduced by mistake, and often in spite of heavy screens of quarantine. But most mammals (other than rats and mice), birds, frogs, toads, and fish have been brought intentionally in the first instances, though many of them have become extremely harmful or unpopular afterwards. It would need a long review to trace all the histories of these changes, and perhaps what has been said about the muskrat will do well enough for a typical pattern of events. A recent monograph brings much of the history up to date for mammals—200 species of them! But most of the introductions failed or did not explode in earnest.[73] Perhaps the following mammals have been the most explosive in various countries to which they have been brought: from the Palaearctic Region, the rabbit, European hare, fox, fallow deer, red deer, and Japanese deer; from the Nearctic, the grey squirrel and muskrat; from the Neotropical, the coypu; from the Oriental Region the mongoose and axis deer. Australia does not seem to have made contributions that matter, except to New Zealand, and New Zealand is the most special of special cases—a land totally lacking native mammals other than bats. The modern meeting there of some Australian wallabies and the brush-tail opossum with placental mammals of various kinds, in a way recapitulates the mixing of faunas in the Eocene over Europe and North America, and in the Pliocene across the Isthmus of Panama. The opossum at any rate is doing very nicely. We usually hear most of all about the spread of the rabbit; but the European hare, *Lepus europaeus*, is now as cosmopolitan as any other truly wild mammal, with bases in Ontario, Brazil, Argentine, Australia, and New Zealand. Of course other species of the genus *Lepus* already occupy all the continents except Australia, though South America only since late Tertiary times.

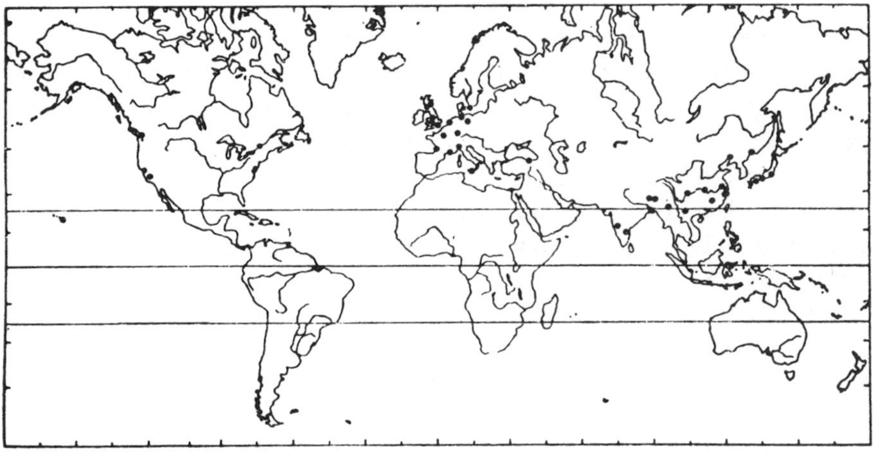

FIG. 31. Distribution of the moth *Aphomia gularis*, that has spread from the Orient across the Northern but probably not yet to the Southern Hemisphere. (From K. G. Smith, 1956, by permission of the Controller, H.M. Stationery Office. Crown copyright.)

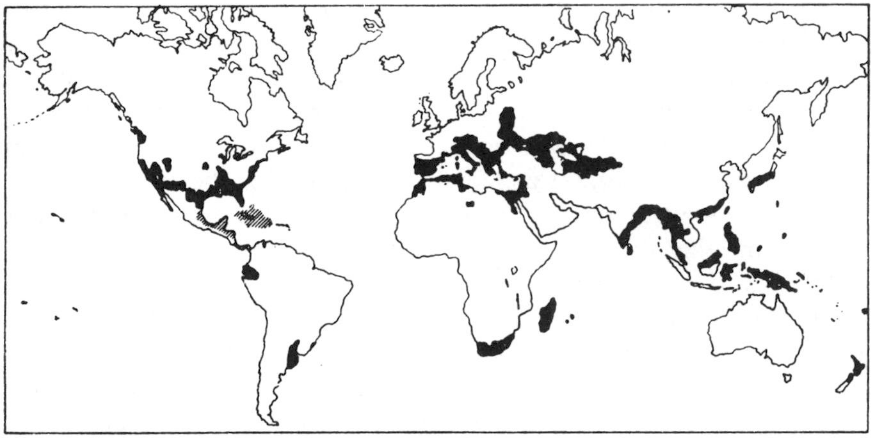

FIG. 32. World distribution of the topminnow or mosquitofish, *Gambusia affinis* (shown in black), whose native home is in south-eastern North America. The cross-hatching shows the native range of other species of this genus. (After L. A. Krumholz, 1948.)

The best history compiled about the spread of birds in any continent was done by Phillips for North America.[97] Up to 1927 only a few species had managed to become permanent invaders on any scale. Many failed entirely, such as thrushes, several finches and titmice, nightingale, woodlark, robin, dipper, corncrake, capercaillie, and mute swan. Some spread well and seemed all right, but then practically died out sometimes twenty years afterwards, the goldfinch, *Carduelis carduelis*, and skylark, *Alauda arvensis*, being among these. (In the United States the skylark is extinct.) Some stayed only in the towns, like the rock pigeon, *Columba livia*, which seldom colonized the sea-coast caves and rocks that are its ancestral habitat in Europe. Four kinds of birds stand out as really successful colonists: the house sparrow, *Passer domesticus*; the starling, *Sturnus vulgaris*; the common ('Hungarian') partridge, *Perdix perdix*; and the pheasants, *Phasianus torquatus* from China and *colchicus* from Europe, and their various mixtures. Also from Asia, the crested mynah, *Aethiopsar cristatatellus*, spread from Vancouver since about 1897 into parts of British Columbia.[99] But it could be said of all these birds that their headquarters was in cultivated and urban lands, and we have yet to see any foreign species other than game-birds penetrate American forests in the way that insects have begun to do so devastatingly.

Much cross movement goes on every year with fresh-water fish; but no one seems to have made good maps to show the course of spread, except for the Pacific salmon mentioned in Chapter 5. There are three impulses that have generated and kept these introductions going. First is the one that makes fish an object of outdoor sport or capture for food. This has for instance sent ordinary brown trout and the North American rainbow trout extremely far round the world to places like East Africa and New Zealand. The second had used fish as allies in the control of malaria. *Gambusia affinis*, a small topminnow belonging to the cyprinodont order of fish, is now referred to as the mosquitofish on account of its great appetite for the larvae and pupae of *Anopheles*. Although other small fish have been introduced to malarial countries for the same purpose, this species is outstanding, not only in its performance, but in having become easily acclimatized without special management in so many parts of the world (Fig. 32). The original range of *Gambusia affinis* is in the south-east of North America.

It is now possibly the most widely distributed fresh-water fish in the world.[92]

The third motive has started a quite modern trend. Thousands of people now keep aquaria in their houses, or even shops and restaurants, with tiny brilliantly coloured tropical fish. Hundreds of kinds of tropical fish are drawn into this growing trade, and Myers has pointed out that some are escaping and invading natural waters as well.[93] 'Word-of-mouth reports have it that the Chinese *Macropodus opercularis*, and the Mexican *Platypoecilus maculatus* and *Xiphophorus hellerii* are breeding in the Everglades', that is, in Florida.

CHAPTER FOUR

The Fate of Remote Islands

When Captain Cook anchored off Easter Island in March 1774, he noted that 'Nature has been exceedingly sparing of her favours to this spot'.[122] This was exactly true, for Nature had only with great difficulty managed to get there at all. The nearest continent is South America, 2,280 miles away, and even the nearest vegetated Pacific island (Ducie Island) a thousand miles. This bit of volcanic rock (from which the famous hatted statues were carved out), covered with hills and grassy downlands, is only about a third the size of the Isle of Wight (Pl. 23). This is about 9 per cent. of the combined areas of the Marquesas Islands—one of the remotest of the Pacific mountain island archipelagos; these in turn are about 8 per cent. of the combined area of the Hawaiian Islands (which amounts to 6,400 square miles). The Hawaiian group is the largest, most varied and richest in life of the truly oceanic islands of the central Pacific: the area of Africa is nearly twelve million square miles!

Plants and animals have managed not only to reach these remote archipelagos and islands without the help of man, but in some have evolved luxuriant tropical vegetation and sometimes, though not very often, unique and peculiar groups of plants and animals. The island of Krakatau, which blew off its head in 1883 and absolutely destroyed all life under a rain of hot volcanic ash that lay more than a hundred feet deep on some of the slopes, was recolonized by plants and animals from the nearest land, and after fifty years had already a rich and maturing jungle of forest inhabited by epiphytic plants and many kinds of animals. By 1933 there were at least 720 species of insects, 30 kinds of resident birds, and a few species of reptiles and mammals, though no frogs or toads. But these species only had to cross by various means from the adjacent tropical lands of Java and Sumatra, a mere twenty-five miles over the sea.[123]

When the first white man made collections there Easter Island had extremely few native plants and animals compared with Krakatau, though this has only 12 per cent. of the area of the former. The Swedish Expedition under Skottsberg that visited Easter Island for a short time in 1917 has published a very good series of reports on the place, as well as upon Juan Fernandez—Robinson Crusoe's island.[144-6] But there are two things that have to be considered besides the remoteness of the island. One is that it has been a great deal modified by human activities, especially grazing sheep and cattle and the removal of timber (Pl. 24). It seems likely that the original condition was a sort of forest savannah with grass. So some of the indigenous plants and animals may have died out before they could be collected by biologists. The only tree, *Sophora toromiro*, is nearly extinct now. The second thing is that no absolutely complete collection of insects and other small animals has been done, and even the Swedish party only spent a fortnight there. Nevertheless, there are certainly no native earthworms at all, only one introduced species; and no land birds or other vertebrates except a few introduced by man. In the flora there are 31 species of flowering plants, apart from cultivated plants like the plantain, sweet potato, and sugar-cane; 15 kinds of fern, of which four are endemic; 14 kinds of moss, of which nine are endemic. That is, less than fifty species that may originally have been native to the island—a tiny flora. The number of animal species that seem to be native is almost absurd. There are so far known to be only five endemic: a green lacewing,[139] a fly,[125] a weevil,[113] a water beetle, and a land snail (Pl. 25).[135] The water beetle, *Bidessus skottsbergi*, was found among algae in the crater lake of Rano Kao, where there are also some kinds of endemic aquatic mosses.[157] No other fresh-water animals have yet been found, though probably microscopic life would be rich enough, because it is easily air-borne even to distant lands. Practically all the rest of the land animals are either known to have been introduced, or else this can be supposed from their cosmopolitan man-borne distribution: about 44 kinds of insects, spiders, and other invertebrates, of which one (a dragonfly) probably arrived under its own power; two introduced lizards; two kinds of birds brought from Chile; and rats. The surviving native animals are therefore outnumbered in species by about 10 to 1, and far more so in populations.

There are thousands and thousands of small remote islands that have, like Easter Island, been too far from the busy evolutionary centres of the continents to acquire more than a sprinkling of accidental immigrants before the arrival of man began to make this process of dispersal so much easier and faster. The small atoll of Palmyra Island in the equatorial Pacific had only fourteen species of native plants. The insect fauna of Midway Island, lying at the western extremity of the Hawaiian chain, was also minute: of beetles there were only six species, of flies only nine.[128]

Before considering some of the more catastrophic invasions of oceanic islands, it is worth examining what is happening on three very remote islands in the South Atlantic—the Tristan da Cunha group. Here a very thorough biological survey was made by the Norwegian Scientific Expedition to Tristan da Cunha in 1937–8, under the leadership of Christophersen.[121] There are also earlier records, especially for the plants and birds. These three islands (Tristan, Nightingale, and Inaccessible) are even farther away from the nearest continent than Easter Island—2,900 miles from South Africa; 3,200 from Brazil and 4,500 from Cape Horn. Tristan itself is the upper part of a volcano risen over 12,000 feet from the sea bottom, and having about half of this exposed above the sea. On the top is an ancient crater with a lake inside it. Down the sides there grows fairly rich vegetation, with only one kind of small tree but with tree ferns, and there is much heavy tussock grass and rather wet heath. On a shelf of land above the shore three miles long lives the small human community.

There are some fifty native and seventy alien species of flowering plants on these islands, besides nearly 300 of ferns, mosses, and liverworts.[120] Animal life, other than sea-birds, is very poor. Five species of birds, some of which have evolved differences on the separate islands—a member of the flycatcher family that looks like a thrush, two kinds of finch, a flightless rail and coot (the rail on Inaccessible Island, the coot on Tristan Island, though now almost extinct).[129] The insect life can be illustrated by some examples. Of the twelve recorded moths and butterflies, the Expedition took only eight. Only five of the twelve appear of be native, the rest brought by man.[150] Only four kinds of plant-lice, of which two at least have come in with man.[141] The beetles were analysed with special thoroughness. Of the twenty species thought to be indigenous, only two are predatory and the rest herbivorous, many of the latter being weevils which are one of the widespread kinds of beetle in oceanic islands elsewhere (the fifteen Tristan ones are all peculiar, and are flightless). Besides these native beetles there are six or seven that are or seem to have been brought in by man.[116]

Consider how extremely little traffic has gone between Tristan da Cunha and other places. Yet the fauna will soon contain as many invaders as there used to be native fauna. In 1882 a ship was wrecked and a few rats got ashore from it. The pastor strongly urged that these should be destroyed, but they were allowed to get in, and now infest many parts of the island of Tristan, eating potatoes (the people's most important crop on land) and killing nesting birds in the wilder parts of the mountain.[114] They are supposed to have destroyed the Tristan coot, perhaps assisted by feral cats, and it is fortunate that rats have not reached the two other islands yet. There are no records so far of great outbreaks among the introduced insect populations, in fact the chief enemy of the potato is a native moth: it has no parasites at all. But one of the invading plant-lice, *Myzus persicae*, is able in other countries to carry two of the worst potato viruses, and one of the moths is a well-known eater of cruciferous plants. It is to be noticed that some introduced species are still confined almost entirely to the limited shelf of settlement, with its pasture and gardens and potato crops; but that others like the rat have spread to the natural habitats as well. An introduced staphylinid beetle, *Quedius mesomelinus*, and a species of European millipede, *Cylindroiulus latestriatus* (that has also been spread by man to North America, South Africa, and the Azores), has colonized a very wide range in tree-fern ground, bogs, the sea-shore drift line and other places. But another European millipede, *Blaniulus guttulatus*, was found on cultivated land.[130]

Here then is an oceanic island in which man has carved out a small patch for himself, leaving the rest of it wild. Except for exploitation of wood and of seabirds, his new influence in the wilder parts is through invading species brought on ships. To see the same process acting on a much larger island group, we may turn to Hawaii.

No need to describe the Hawaiian Islands: remote, mountainous, volcanic, tropical, rich, and until modern times holding within their archipelago one of the most extraordinary island floras and faunas ever known. Few people now deny that these islands are truly oceanic, like Tristan da Cunha and Easter Island, and that the endemic species there are descended from rare immigrants that had to cross several thousand miles of ocean. America and Australia are over four thousand miles away; the nearest continental islands are Japan—3,400 miles. Fiji—near or on the edge of the old sunken outlier of the Australasian continent—is about 2,800 miles away. Before the Polynesian canoes reached Hawaii in about the twelfth century A.D. or earlier the islands were probably covered with luxuriant forest, except in places where the climate is locally dry or there were recent lava flows. Since then the forest line has retreated from the coast until it now covers only a quarter of its former extent, on the mountains mostly. Fire and wild cattle, sheep, goats and horses, and the clearing of land for crops, have all contributed to this retreat. But from a quite rich percentage of surviving forms and the earlier records a pretty good stock-taking has been made, though many species may have died out before white men came, and it has even been suggested that as many as a third of the original insect fauna had disappeared unrecorded. Over nine-tenths of the 1,729 species of flowering plants are found nowhere outside these islands. Zimmerman, to whose *Insects of Hawaii* I am indebted for much critical information, estimates that 3,722 of the 6,000 or so species of insects known there are also endemic; the rest being comprised of species also living elsewhere, naturally or artificially introduced.[158] There are two large families of land snails. The Achatinellidae with 215 species are unique to the islands; the Amastridae with 294 almost so (Pl. 26). The former family lives entirely in trees. The shells are gaily marked and coloured. The Amastridae show a great deal of evolution into different ecological forms, both in trees and on the ground. Of the 77 kinds of endemic birds ever found on the whole Hawaiian chain, about 43 species and sub-species belong to the Drepaniidae, a family within which more than a dozen ecological ways of life have been evolved within Hawaii—honey-suckers, wood-insect hunters, other insect-eaters, seed-eaters, nut-eaters, fruit-eaters—differences that would in a continent be developed in separate orders, not just genera of birds. It is not surprising that Captain King was puzzled when he saw one. In 1779 he wrote: 'A bird with a yellow head, which, from the structure of its beak, we called a perroquet, is likewise very common. It, however, by no means belongs to that tribe, but greatly resembles the yellow cross-bill, *Loxia flavicans* of Linnaeus.'[131] (Pl. 27). This was the Drepanid *Psittacirostra psittacea* (Pl. 27). Taxonomists have tried to calculate how many ancestors these Hawaiian groups may have had; that is, how many original ancestors arriving by various routes to the islands. For flowering plants it is about 272 species; for insects between 233 and 254; for Achatinellidae only one; for land snails from 22 to 24;[158] and for birds about 14—the Drepaniidae coming only from one of these forms.[133]

What has been the fate of this marvellous flora and fauna? First of all the list has been enormously added to by introduction, partly on purpose and partly by mistake. The full roll-call for insects has not yet been finished, but out of the 1,100 or so species given in the first five volumes of the *Insects of Hawaii*, 420 are thought to be adventive. This is a rough estimation and there still are nine orders of insects to be assessed, including such predominantly important ones as moths, beetles, flies, and Hymenoptera. In 1953 a list was published of 49 'economic' insects found to have become established since 1939.[137] The geographical sources of these species are very mixed. They came from California, Mexico, the Philippines, Samoa, Fiji, Guam, Saipan, and New Guinea, and for some the origin is unknown. Among these immigrants was the Argentine ant (probably from California) in 1940. When Wheeler compiled a list of Hawaiian ants in 1934 he mentioned that this species had been intercepted by quarantine and had not by then invaded the islands.[153] He also recorded that the leaf-cutting ant *Pheidole megacephala* was then displacing a previously introduced ant, *Solenopsis rufa*, on the island of Oahu. Zimmerman wrote in 1948: 'The voracious *Pheidole megacephala* alone has accounted for untold slaughter. One can find few endemic insects within the range of that scourge of native insect life. It is almost ubiquitous from the seashore to the beginnings of damp forest. Below about 2,000 feet few native insects can be found today.' It was known that the leaf-cutting species had invaded the Canary Islands and Madeira, to be followed at a later date by the Argentine ant, which not only wiped out *Pheidole* but also practically all the native ants below 3,000 feet.[153] Perhaps the same thing will now happen in Hawaii.

Every new insect pest may cause a train of operations with foreign counterpests. A very recent report on the annual increment of counterpests to Hawaii in 1953–5 gives quite a vivid notion of this process.[152] For control of the shrub *Lantana*, a Mexican longicorn beetle that bores in the stems, also a Central American chrysomelid beetle and a phalaenid moth from California whose grubs and caterpillars respectively eat the leaves. There are already an introduced seed-eating fly, and some other insects for this job. Then from Mexico a fly whose larvae eat the flower heads of *Eupatorium glandulosum*, a relative of our own hemp agrimony that has become a tropical weed in Hawaii, as well as in the Philippines and elsewhere. A moth from Brazil to eat the leaves of another locally troublesome plant, the Christmas berry tree. Four kinds of Mexican dung-beetles, whose grubs might help in controlling the maggots of some kinds of flies. Two Scoliid wasps from Guam, to attack various kinds of scarabaeid beetles. Two parasites from Arizona to try again (after a failure to establish them ten years earlier) in the control of a moth that attacks the flowers of the mesquite tree. A Mexican ladybird to feed on aphids in the sugar-cane fields. Finally, two carnivorous snails, one from the Mariana Islands and one from Florida, to try against giant African snails. It is quite an exchange and bazaar for species, a scrambling together of forms from the continents and islands of the world, a very rapid and efficient breaking down of Wallace's Realms and Wallace's *Island Life*!

Most of the herbivorous insects have followed in the wake of earlier plant introductions (as field crops, fruit trees, forest trees and garden plants) and it is usually some years before the animals catch up with their plant hosts, as has been seen recently in the case of *Leucaena glauca*.[143] This large leguminous shrub is probably an original native of South America, though it has been spread to other parts of the world, including Hawaii since 1888. It is valued there as a forage crop that is full of protein, and its seeds were collected and sown, and also later on became used in the island manufacture of seed jewellery. It became a thriving additional crop on the islands, but in 1954 a small anthribid beetle from the region of Indo-China and the Philippines was discovered to be living in the seeds and, on the island of Oahu, sometimes destroying the complete seed crop. Hitherto its control has not been achieved.

The native life is not just retreating with the forest, keeping its forces intact though on a smaller area. It is true that a good deal of the forest, and of some other upland habitats, survives because it is impossible to cultivate. Yet roaming cattle and other feral animals have done much harm. And ship rats, whose violent influence is a frequent refrain in the modern history of islands, have also gone into the forest. The native moths have diminished greatly from their former strength and some have died out. This Zimmerman attributes to the invasion of forest by ichneumonid parasites brought in as counterpests on agricultural land, and he cites especially three species, *Casinaria infesta*, *Cremastus flavoorbitalis*, and *Hyposoter exiguae*, that have a very wide range of hosts. For Hawaiian insects were not naturally parasitized or adapted against parasites and insect enemies to the extent that continental insects are. Furthermore, this decrease in native moths may be the reason why some species of *Odynerus*, a genus of hunting wasps that has many species in Hawaii, have also declined in numbers; for they depend on caterpillars for stocking larders in which their own young grow up.

Amongst the many invaders of Hawaii none can have had such a long and steady progress across the Indo-Pacific world before its arrival as the giant snail *Achatina fulica*. This genus is otherwise entirely Ethiopian in distribution, with over 65 species that live chiefly in tropical forests. It contains the largest living land snails, the biggest of all, *Achatina achatina*, being still confined to West Africa, where it is a favourite food of the natives. *A. fulica*, though rather smaller (Pl. 28), is something to be considered if you have to collect 400 of them every night in a small garden, as a resident of Batavia in Java was doing in 1939.[78] This is an East African species that may have been introduced to Madagascar long ago. It began to spread to the outlying islands of Mauritius (by 1800), Reunion (by 1821), the Seychelles (by 1840) and the Comoro Islands (by 1860). Some were released in Calcutta in 1847, and Bequaert, who has documented its travels, as well as the systematics of the whole group, says that 'at first, the spread of the snail in southern Asia was very slow'.[115] It was in Ceylon by 1900, Malay Peninsula certainly by 1922 and probably twelve years earlier; Borneo by 1928; Siam in 1937–8; and Hong Kong in 1941. It moved through the Netherlands East Indies in the nineteen-twenties and thirties. It was in Japan, though not doing very successfully, by 1925, reached the Palau Islands in 1938 and on to the Marianas, soon becoming a major agricultural problem in many of the Micronesian islands. In Guam especially it became a plague, having been brought there from Saipan in the Marianas in 1946. By 1948 it had a foothold at three points in and near New Guinea. Meanwhile a few got into California about 1947; but it is not thought likely that it can ever become established in the United States, because the climate is unsuitable for a tropical snail. This majestic spread was accomplished by partly accidental dispersal on transported stores and plant materials; and also to quite a large extent because of its value for food. Their size, voracity, and abundance give a remarkable atmosphere to these invasions. There can be few invading species which become such a menace to motor traffic that they cause cars to skid on the roads! The whole extended snail may measure nine inches, not counting the projection of its shell.

The giant snail reached Hawaii in 1936, and in spite of great efforts for its control, it now inhabits Oahu where it damages crops—an invasion thought to have started from only two individuals brought from Formosa. The trial of enemy snails to kill *Achatina* is already in full swing, and it may be noted that these have been brought not only from the original home of the species in East Africa, but also from one of the Mariana Islands and from Florida.[154] Meanwhile many of the beautiful native Amastrid snails have become scarce in the lowlands and some species extinct. One factor bringing this about seems to have been the attacks of foreign rats (Pl. 26);[147] the original Polynesian rat having rather different food habits, cannot have been decisive in causing this decline.

The birds also display the consolidation of alien species and, on the whole, the diminution of native ones. In 1940, when E. H. Bryan summarized the position in Hawaii, there were already about half as many introduced as original native forms: 94 kinds of foreign birds had been tried, and only 41 found wanting.[117] The 53 established ones show the varied pattern of origins that is becoming familiar in this book. The ring-necked pheasant, *Phasianus colchicus*, derived from Europe; the green Japanese pheasant, *Phasianus versicolor*; the California quail, *Lophortyx californica*; the painted quail, *Coturnix coturnix*, from Japan; the lace-necked dove, *Streptopelia chiensis*, from Eastern Asia; the barred dove, *Geopelia striata*, from the Malay Archipelago—to mention only some game-birds that have now got a firm hold in the islands.[142] The two pheasants have not only spread widely, but in some places hybridized (Pl. 29). The wild jungle fowl, *Gallus gallus*, must have been brought from Malaya by the Polynesian voyagers themselves, as these birds existed in Hawaii when Captain Cook discovered the islands. The rock pigeon, *Columbia livia*, instead of being only a town bird as it is in the United States, has become wild on the cliffs. The Indian mynah, *Acridotheres tristis*, brought from India in 1865, is well known to ecologists because of the part it played in originally spreading the seeds of *Lantana*.

According to Munro, who had known the Hawaiian birds for a lifetime, and was with Perkins and other early naturalists when they explored and collected at the end of the nineteenth century, there are several introduced birds that have penetrated more or less deeply into the forests.[134] The babbler or Pekin nightingale, *Leiothrix lutea*, a Chinese species brought over in 1918–20, is now on most of the main islands, and it is on record that bird malaria has been found in this species in Japan. There are also the Chinese thrush, *Trochalopterum canorum*, from about 1900, a bird of the scrub layer that has gone deeper into the forests than any other species, but in 1944 was reported to be diminishing locally; and the Japanese tit, *Parus varius*, from 1890 onwards, which has made itself quite at home in the forest. It has been suggested, though without direct proof, that species like these might carry diseases of birds from the lowlands into the upper zones, and possibly harm the Drepanids. These wonderful birds have practically all become reduced in numbers, even in their remaining natural haunts. Only about a third of the species and island sub-species have a good chance of survival in the future. Some which were thought probably to be extinct have yet been found to persist, but have been missed through the physical difficulty of searching for them:

thus the crested honey-eater, *Palmeria dolei*, had not been seen since 1907, yet was telephotographed on the high mountains of Maui as recently as 1950. And *Pseudonestor xanthophrys*, not seen for half a century, was observed on Maui at 6,400 feet in 1950.[140]

Captain King's 'perroquet', the Ou, *Psittacirostra psittacea*, used to be abundant on the main islands, but Perkins said in 1903 that though it was still widespread on other islands the bird had become practically extinct on Oahu. This he thought might be caused by competition from the ship rats, *Rattus rattus*, that had spread in the forests there: 'Now over extensive areas it is often difficult to find a single red Ieie fruit, which the foreign rats have more or less eaten and befouled, and they may thus have indirectly brought about the extinction of the Ou.'[138] The bill of this bird is indeed especially useful for picking out the fruits of the Ieie, a bright-flowered liana, *Freycinetia arnotti*, that climbs on forest trees; and the Ou's chief habitat on all these islands was in the zone where the liana occurred. But the birds have decreased also on the other islands, they will eat other fruits and also caterpillars (on which they feed their young), and have for some time been known to feed on the fruits of introduced plants like the guava, which itself is a wide-spreading invader in Hawaii. It is also known that domestic birds have brought in diseases; Drepanids have been seen with at least one of these; and bird malaria has come in to the islands, and may possibly be spread by mosquitoes.[112] The mosquitoes are apparently all introduced species, the islands originally being free from them. W. A. Bryan in his *Natural History of Hawaii* says that the one that bites at night, *Culex fatigans*, was thought to have arrived on a ship from Mexico as early as 1826. There are also some day-biting species of *Stegomyia*.[118] All these possibilities only serve to suggest the tangle of influences that are likely to be at work: no one has sorted them out in a thorough way. For example, what of the scarcity of native caterpillars affecting young forest birds? The Ou still survives, though in 1944 it was thought to be dangerously near extinction. In 1950–1 some were seen in a mountain forest reserve and in the National Park on the island of Hawaii, at a height of several thousand feet. It mitigates the fate of remote islands in this century if some of their species are saved from the wreckage, like the few survivors that clamber out of

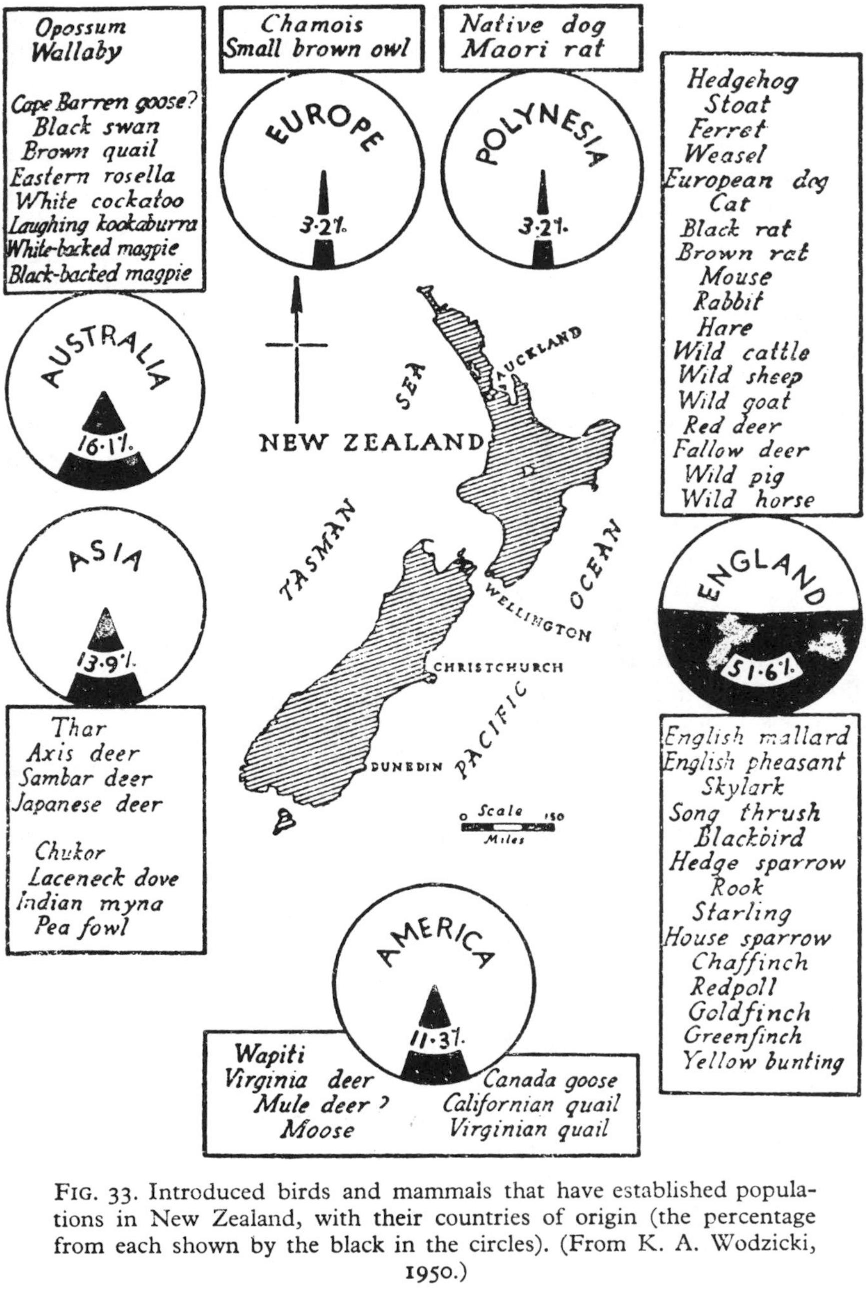

FIG. 33. Introduced birds and mammals that have established populations in New Zealand, with their countries of origin (the percentage from each shown by the black in the circles). (From K. A. Wodzicki, 1950.)

a smashed aircraft. This bird especially, of which Perkins wrote: 'Sometimes it sings as it flies, and when a small company are on the wing together they not infrequently sing in concert, as they sometimes do at other times, and in a very pleasing manner.' [138]

The only rival to Hawaii among remote islands is New Zealand. Here is a country that looks like part of a continent, yet was probably never joined to one directly, and has been isolated for an immensely long period. It has therefore partly the environment of a continent but the history of an oceanic island. No place in the world has received for such a long time such a steady stream of aggressive invaders, especially among the mammals—successful in the short run, though often affecting the future of their own habitats in a decisive manner. Originally there were no native mammals except bats. In 1950 Wodzicki could write an ecological monograph upon some twenty-nine kinds of 'problem animals', among them being eleven kinds of ungulates: four from Japan, four from North America and three from Europe (including England), and to these must be added feral (and domestic stock), not neglecting pigs.[156] A list of such successfully established mammals and birds compiled by Wodzicki is set out in Fig. 33. Red deer, *Cervus elaphus*, were liberated between 1851 and 1910, and quickly multiplied in both islands of New Zealand (Pl. 30). Now spreading patches from different centres have merged within each Island, but the greatest occupation is still in the South Island (Figs. 34–5). The red deer have already made a profound impact upon native forests, especially in the drier types of woodland; but it is in the wetter regions that forest damage leads to most serious soil erosion. It is likely that on many watersheds the deer, helped by domestic stock, have tipped the scale towards a cycle of catastrophic soil erosion, which is felt not only in the mountains but also in those parts of the lowland valleys that receive the extra load of silt washed from above.

To follow the story of invading insects in New Zealand would only repeat what has already been indicated for other countries. Taking only populations that had arrived and become noticeable, there were fourteen species between 1929 and 1939, and another nine by 1949.[124] Among the last was the common ground-nesting wasp of Europe, *Vespula germanica.*[136] In 1945 beekeepers at one place in the North Island observed some strange

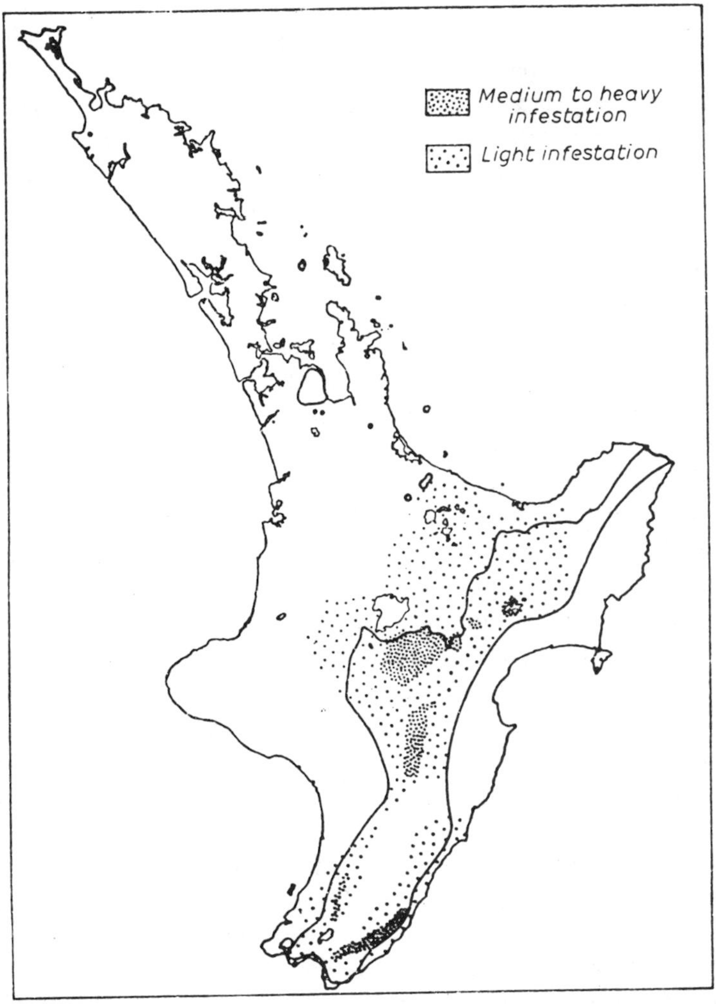

FIG. 34. Areas occupied by the introduced red deer, *Cervus elaphus*, in the North Island of New Zealand, 1947. The southern beech, *Nothofagus*, forests are enclosed by the black line. (From K. A. Wodzicki, 1950. Forest areas mapped by C. M. Smith and A. L. Poole.)

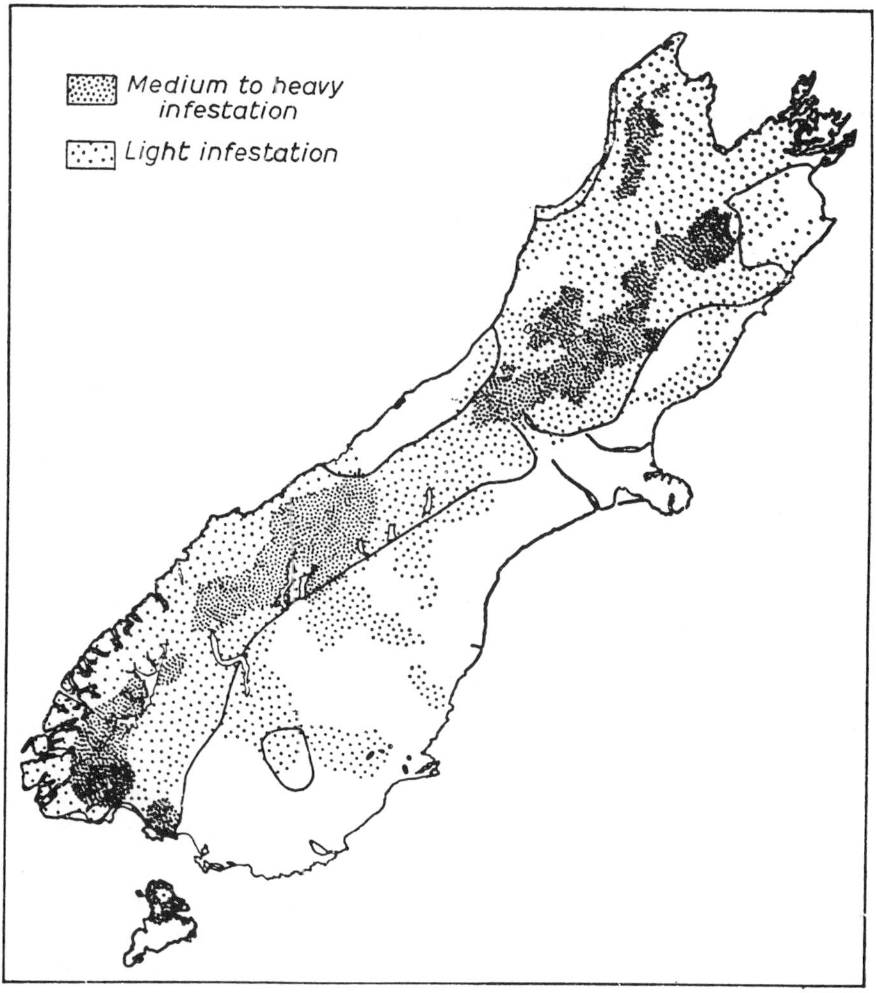

FIG. 35. Areas occupied by the introduced red deer, *Cervus elaphus*, in the South Island of New Zealand, 1947. The southern beech, *Nothofagus*, forests are enclosed by the black line. (From K. A. Wodzicki, 1950. Forest areas mapped by C. M. Smith and A. L. Poole.)

wasps flying about their hives. After this date the wasps spread and increased at a great rate, probably aided by the absence of winter cold to check their breeding. That year seven nests were destroyed, in 1946, 140, and by 1948 over 3,000—and this did not stop the spread. When a bounty was offered for resting queens, one schoolboy brought in 2,400 in less than a week!

The fate of remote islands is rather melancholy, even after one has made allowances for all the human excellence that has remained or developed again in some of them after our invading civilizations settled down. The reconstitution of their vegetation and fauna into a balanced network of species will take a great many years. So far, no one has even tried to visualize what the end will be. What is the full ecosystem on a place like Guam or Kauai or Easter Island? How many species can get along together in one place? What is the nature of the balance amongst them? Can we combine the simple culture of crops with the natural complexity of nature, especially when there is an almost inexhaustible reservoir of continental species that may send new colonists to disturb the scene? All these questions are much nearer than the horizon, though most ecologists have not looked at them with any enthusiasm, or if they have glanced at them, shuddered and turned away towards the already tedious and difficult task of understanding the biology of a single species, dead or alive.

I would like, however, to leave the subject with a back-glance at a more pleasant and balanced ecological world, before Atlantic civilized man crashed into this remote galaxy of island communities. In that age, when the numbers of human beings were regulated by customs, often harsh enough, but meeting the end desired, a great many of the Pacific Islands were inhabited by quite large numbers of a small species of rat, derived from a Malayan form, and evidently brought by the Polynesians in their great migrations eastwards and southwards some hundreds of years ago.[149] *Rattus exulans* (with closely similar forms like *Rattus hawaiiensis* (Pl. 31)) is a small rat, much gentler in habits and less aggressive than the larger ship and Norway rats: it has been found in New Zealand for example that the Maori rat does not do harm in bird island sanctuaries. On many of the islands of the Pacific the native rat was exterminated either by cats or by the arrival of these larger species. For a long time it was believed that they were extinct in Hawaii, until it was discovered that they had been confused with the young of the grey form of the ship rat, and are actually living there with the other two foreign species of rat.[148] From early missionary books we learn that these little rats were an important part of civilized life. On the island of Raratonga, in Mid-Pacific, they were highly prized for sport. 'In those days—ere the cat had been introduced—rats were very plentiful. Rat hunting was the grave employment of bearded men, the flesh being regarded as delicious.' [127] And on the Tonga Islands about 1806, the King and court used to go out and shoot the rats along the forest paths, using huge bows and arrows six feet long: 'Whichever party kills ten rats first, wins the game. If there be plenty of rats, they generally play three or four games.'[132] There were elaborate rules, as we should now have for football or hunting: precedence, offside, and—a wonderful dispensation—if you shot a bird you could count it as a rat! Even late in the nineteenth century 'the proverb "sweet as a rat" survives in Mangaia'.[126] Von Hochstetter, writing in 1867 about the small Maori rat in New Zealand, relates that 'this indigenous rat was so scarce already at the time of the arrival of the first Europeans, that a chief, on observing the large European rats on board one of the vessels, entreated the captain to let these rats run ashore, and thus enable the raising of some new and larger game'.[151] Returning to Captain Cook at Easter Island in 1774: 'They also have rats, which it seems they eat; for I saw a man with some dead ones in his hand, and he seemed unwilling to part with them, giving me to understand they were for food.' [122] Can we still find a remote island where people will be unwilling to part with the new rats that have arrived there in the last 180 years? Perhaps we could bear in mind the story told by Buxton and Hopkins, about the arrival of the human flea in one small Pacific Island more than a hundred years ago: 'The placid natives of Aitutaki, observing that the little creatures were constantly restless and inquisitive, and even at times irritating, drew the reasonable inference that they were the souls of deceased white men.' [119] We may hope that this same restless curiosity in the form of research will find out how the broken balance can be restored and protected.

CHAPTER FIVE

Changes in the Sea

For though I scorn Oceanus's lore,
Much pain have I for more than loss of realms:
The days of peace and slumberous calm are fled;

That was before we knew the winged thing,
Victory, might be lost, or might be won.
Keats, *Hyperion*

In contrast to land and fresh waters the sea seems still almost inviolate. Yet big changes in the distribution of species have already begun as a result of human actions during the last hundred years. These actions are of three kinds. First the digging of new canals. Secondly, accidental transport on ships. And thirdly, deliberate introductions. The Panama Canal, though it has in a formal sense split the Nearctic from the Neotropical Region once more, is hardly a serious gap, nor much of a transport line for marine life from one ocean to the other. In 1935 and 1937 Hildebrand made a survey of the animal life in the locks and inner channels of the Canal and found that a good many fishes and some other animals have moved part of the way into the system from each end.[179] Indeed there is no physical obstacle to prevent them from doing so, and he prints a photograph of men picking up a number of very large fish after the emptying of one of the locks. The real barrier is the forty miles of fresh water, especially the great Gatun Lake. The fish that have penetrated at all are, as one would expect, those that can live in brackish and even in fresh water—various gobies and also other kinds of tropical fish. The only species known to have made a complete crossing is the tarpon, *Tarpon atlanticus*, of which four were found in the lowest lock on the Pacific side when it was emptied in 1937. They have also been reported at the Pacific sea-level terminus, but had not (in 1939) been caught at sea in Panama Bay. They seem to be quite frequent in Gatun Lake.

The Suez Canal is quite a different matter, though it also presents some serious obstacles to the transit of marine species. Here the reason is the opposite from that in the Panama Canal. The Suez Canal is about 100 miles long, and in the middle there is a stretch of nearly 14 miles of the Great Bitter Lake. The Lake has very high salinity from the dissolving of rock salt deposits laid down in a much earlier period. But, according to Munro Fox who took the Cambridge Expedition to the Suez Canal in 1924, the salinity has grown less than it was, by the mixing with ocean water, and was then still falling.[176] As explained in Chapter 2, the great branch of the Tethys Sea connecting the Mediterranean region with the Indian ocean was severed by Miocene times, and great differences began to appear in the fossil faunas to east and west. The Indian Ocean kept its luxuriant fauna. The Mediterranean became much impoverished, no doubt chiefly because it was already part of the great brackish Sarmatian, and later the Pontian, seas that enveloped much of Central Europe, the Black Sea, and Caspian–Aral region. The detailed history of the Gulf of Suez is complicated, and not yet quite fully elucidated.[186] It is known however that it was for a certain time joined to the Red Sea, because sea-urchins and other fossils from there have been found in its Mid-Pliocene deposits. It also seems certain that it was cut off from the east during all or a great part of the Quaternary Period following this. In modern times the fauna of the Mediterranean and of the Red Sea were quite distinct, indeed they had and still have relatively few species in common. The other canal (from the Red Sea) that the Egyptian Pharaohs built several thousand years ago, could not have provided a highway for marine species, because it had such a long fresh-water stretch, and carried no traffic directly to the Mediterranean.[176]

Since the Suez Canal was opened in 1869 a fairly strong contingent of animals has managed to pass from the Red Sea into the Gulf of Suez and spread into the Mediterranean, some of them rather widely.[197] The exchange has gone mainly in this direction because of the set of currents, the tides for most of the year running westwards from the Red Sea end. Thus only two of the sixteen crabs taken by the Expedition in the Canal were Mediterranean ones.[164] Though the shipping itself must have enabled a good many of them to run the gauntlet, by speeding up the passage through the Bitter Lake, there also seems to have been direct migration. The arrival of the Red Sea crab *Neptunus pelagicus*, a swimming species, was traced through observations made by the Suez Canal Company staff, whose interest was in fishing it for food.[164] It first began to be numerous in the Canal in 1889–93, reached Port Said by 1898 and four years later was common there. By 1930 it was common also in Palestine. Today it is a staple article of Egyptian food, fished for from Port Said, Alexandria, and Haifa, and it has reached at least as far as Cyprus.[174] *Myrax fugax* is another crab that has had a rather similar history of successful invasion. A crab, *Neptunus sanguinolentus*, and a bat-lobster, *Thenus orientalis*, both from the Red Sea, were detected in Fiume Harbour in Italy in 1896. The Red Sea pearl oyster, *Pinctada vulgaris*, has spread as far afield as Tunis.[176] So in the last ninety years we begin to see the redeployment of the fauna of the Tethys Sea. However, I suppose it is likely that the Bitter Lake, whose salinity is more than twice as high as that usually found in the sea, will prevent a good many plants and animals from getting through, or delay them for a long time.

Accidental carriage in or on shipping, that is in water ballast tanks or on the hull, has been a powerful and steady agency dispersing marine plants and animals about the world, just as it apparently carried the Chinese mitten crab to Europe. In 1946 the larvae of a prawn *Processa aequimana* were detected for the first time in plankton hauls from the southern part of the North Sea, and in 1946–8 the numbers of these increased each year. The adults had not yet been found there. This prawn is known to live in the Red Sea; its larvae have been found in the Suez Canal, and adults at Naples.[192]

The bottoms of ships will quickly get growths of sessile marine algae and animals amongst which more mobile forms can hide and feed: whole communities in this peculiar habitat have been surveyed.[178] Captain Joshua Slocum recounted that while he was sailing across the Atlantic alone in the *Spray*, the fishes and dolphins that had been accompanying him turned aside to go with a large sailing ship that had its bottom much fouled in this way, adding 'Fishes will always follow a foul ship'.[195] These growths must provide a habitat for animals over great distances, and must still do so on many modern boats, in spite of the increased use of chemical anti-fouling treatments. It is known for certain that the slipper-limpet *Crepidula* (referred to later on) grows on the bottoms of ships that have been laid up for some time, and may get spread when these are moved to other stations.[161] The arrival of the diatom *Biddulphia sinensis* from the Indo-Pacific to the North Sea about 1903 is also explained in this sort of way. Its subsequent spread and astronomical multiplication there are summarized by Hardy, who gives excellent pictures of this floating microscopic alga.[177] Its spread is not merely of interest because of dispersal, but because it has become one of the dominant phytoplankton species of part of the North Sea, and has spread also to the Irish Sea and Scandinavian waters.

Shore seaweeds are also being moved from one ocean to another. There is a small and inconspicuous red alga, *Asparagopsis armata*, known also as *Falkenbergia rufolanosa* (Fig. 36), that grows at low tidal levels and is abundant along the south coast of Australia, and lives also in

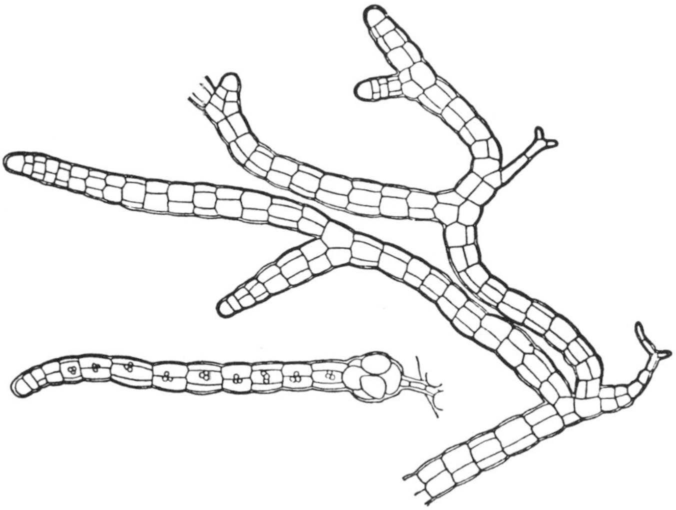

FIG. 36. *Falkenbergia rufolanosa*, an Australasian seaweed recently spread to Europe and North Africa. (From J. and G. Feldman, 1942.)

Tasmania and New Zealand. The exhaustive research of the Feldmans indicates that these two 'species' are alternative life history phases of the same seaweed, *Falkenbergia* being the tetrasporic phase of the other.[175] This conclusion is strongly supported by the recent simultaneous spread of both forms into the Mediterranean and Western Europe, no doubt dispersed on shipping (Fig. 37). *Asparagopsis* was first noticed in the extreme south of the French Atlantic coast in 1923, and in the same year *Falkenbergia* was found at Cherbourg, and *Asparagopsis* in Algeria. The map in Fig. 37 gives the later discoveries up to 1934. In 1941 it was in the West of Ireland, in 1950 well-established in Cornwall, in 1951 in the Scilly Isles,[180] and by 1954 there was a colony in the Isle of Man.[199] There is one other species of *Asparagopsis* that has a world-wide distribution in tropical oceans, but this may be natural.

Elminius modestus is a barnacle that lives on the intertidal rocky shores of New Zealand and Australia. In 1945 it was noticed on the south-east coast of England.[171] It must have arrived at least a few years before this, as a survey in 1947 showed that it was widespread from Norfolk to Dorset, and it was also living in one spot in South Wales. This barnacle is certainly able to get about on the hulls of ships, for it fouls them quite intensely, and was taken early on from a vessel going between Holland and England. It now occupies most of the north coast of France, and lives also in Belgium and Holland.[162] A single individual that had settled on the rocks in 1954 was found on the Isle of Cumbrae in the Clyde, in the course of considerable field research there upon other kinds of barnacles.[170] It has recently been detected also in Cape Town—1949, the first record for South Africa.[193] This is a tough and dominant species, able to occupy the shore in face of competition from other kinds of barnacles, though it does not replace them except in certain zones. It lives chiefly at the lower intertidal levels and below them, flourishing in rather sheltered and muddy waters, thus entering into competition with oysters as well.[167] '*Elminius* ranks as a dominant littoral organism in the estuaries of the Colne, Blackwater, Crouch and Thames.'[171] Other barnacles proved to have crossed the world on the hulls of ships are *Balanus eburneus* from eastern North America to the Mediterranean and thence to Britain; and *Balanus improvisus* from the Northern Hemisphere to Australia.[161]

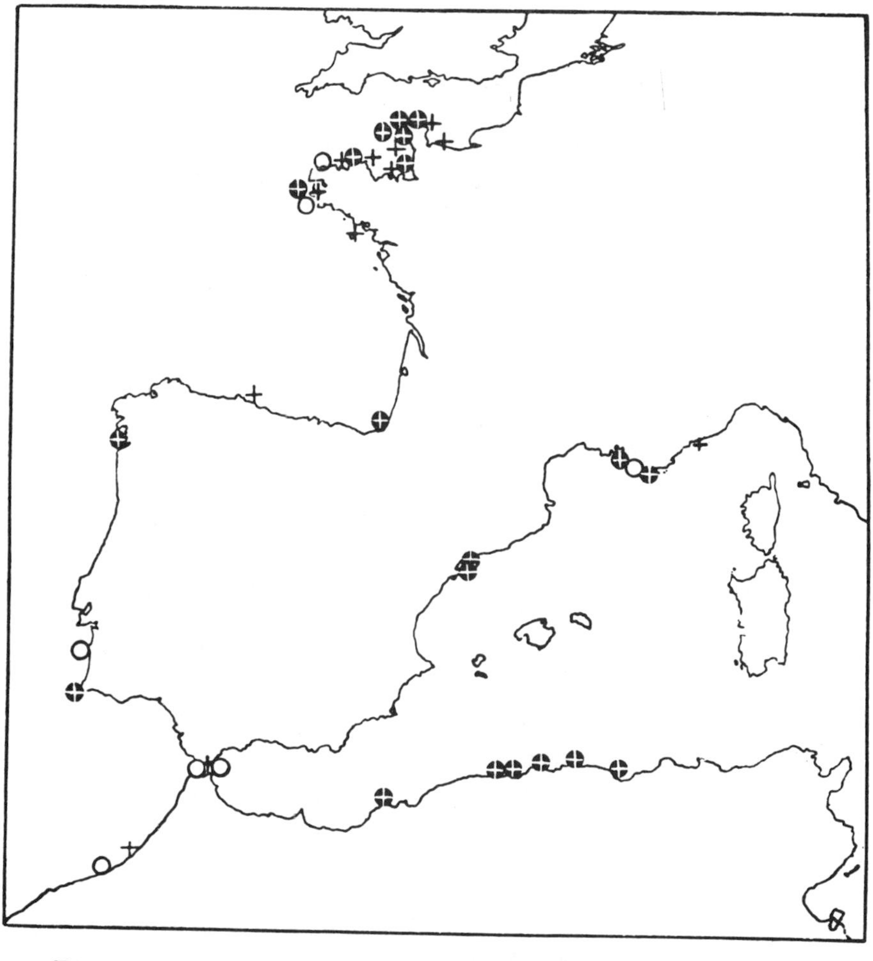

FIG. 37. Simultaneous spread of the two phases of an Australasian seaweed, *Asparagopsis armata* (circles) and *Falkenbergia rufolanosa* (crosses), or both (black with white cross), in south-west Europe and North Africa. (From J. and G. Feldman, 1942.)

But the greatest agency of all that spreads marine animals to new quarters of the world must be the business of oyster culture, a very ancient and world-wide craft now turning gradually into an applied science. It involves much greater managed interference with the natural habitat than any other kind of fishery, and in this way resembles more the crop or flock cultivation of agricultural land, while most other purely sea fisheries still remain at the hunting stage—depending on knowledge and on restraint but not on modification of the habitat in an elaborate way. Two features of oyster culture have deeply affected the spread of species. One is letting the free-swimming oyster spat settle on artificial surfaces like shells, tiles, bamboos, mangrove sticks and the like.[191] These are eventually planted on grounds where the food supply of plankton is rich, to fatten them up for use. The second practice is to bring in foreign oysters and similarly fatten them before they are sold. In England only the native oyster, *Ostrea edulis,* is able to breed and maintain itself. But in the past many shipments have been made of Portuguese oysters, *Ostrea angulata,* and eastern American oysters, *Ostrea virginica,* though these do not establish breeding populations in our waters. An interesting example of the unintentional transport of oysters to a new place by ship was the sinking of a ship at Arcachon in the Bay of Biscay about 1870 with Portuguese oysters on board.[191] This new French colony became one of the regular sources of supply. Oysters are therefore a kind of sessile sheep, that are moved from pasture to pasture in the sea.

The moving about, without particularly stringent precautions, of masses of oysters was bound to spread to other species as well. The first important one was the slipper limpet *Crepidula fornicata,* a native of the east coast of North America, whence it has been transported both to Western Europe and to the Pacific Coast.[200] Its early history in England is not exactly dated, but it first attracted notice at Brightlingsea in Essex about 1890.[167] Since then it has spread along the English coast to Scotland in the east and Cornwall in the west.[166] In 1953 a few were found for the first time in Milford Haven, in the south of Wales.[169] This multiple mollusc, whose individuals sit on top of each other in tiers, has somewhat similar needs to the oyster, since it lives by filtering plankton. It is therefore a serious competitor for space to sit on, especially as it favours the same muddy kinds of shore (Pl. 33). I shall mention this species again in Chapter 6.

A serious enemy of oysters has also come in, though much more recently. Oyster beds all over the world are preyed upon by the small whelk tingles or oyster-drills, of which there are two English species: the dog whelk, *Purpura* (or *Nucella*) *lapillus*, also commonly seen around mussel beds, and known as an important predator of barnacles; and the smooth whelk tingle or oyster-drill, *Ocenebra erinacea*. In 1928 the American oyster-drill or rough whelk tingle *Urosalpinx cinerea* (Pl. 32) was found and has since spread to various oyster beds in Essex and Kent, but not beyond (Fig. 38). It does not have a free-swimming stage and is chiefly moved about by man. We know now that it had probably reached this country in the late nineteenth century.[165] It must be ranked as a really successful invader, living on young oysters as well as other animals, and reaching population densities of five to a square yard. Oyster populations in England have suffered severe disasters in recent decades and can ill afford an additional enemy that is able to destroy half the annual increment of an oyster bed. Oysters are susceptible to very cold winters, and suffered great losses in 1928–9, 1930–40 and 1946–7. *Ocenebra* also declined in numbers and in the latest catastrophe became almost extinct in Essex and Kent, though not on the South coast. But *Urosalpinx*, being less vulnerable to cold, did not decline and so has achieved a dominant place in this community.[167] In 1955 *Ocenebra* was just beginning to reappear in those parts.[188] *Urosalpinx* has also reached the Pacific coast of the United States.[184]

This traffic in oysters and their associates has effects that can only be touched upon in such a short essay as this. In 1949 consignments of *Ostrea edulis* were planted on the American coast in Maine, and began to breed with some promise of permanent populations.[184] The Japanese oyster, *Ostrea gigas*, was first brought over to the coast of the State of Washington in 1905, and in much later years other plantings were made in British Columbia, Oregon, and California, and a great new market for 'Pacific oysters' grew up.[183] But still the spat is grown in Japan, brought over and planted in America, as they only breed sporadically in their new habitat. As usual, other species have come in with the stock:

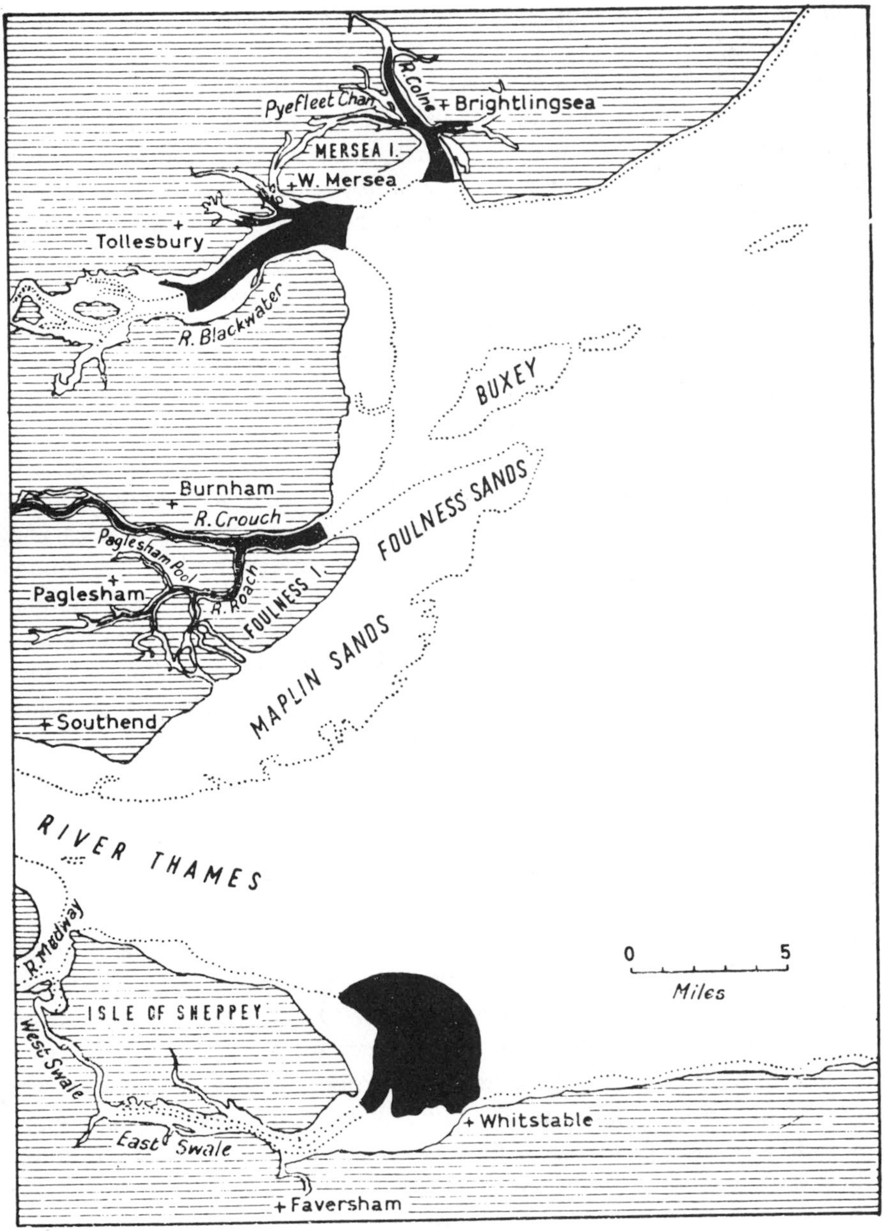

FIG. 38. Known distribution of the American whelk tingle or oyster drill, *Urosalpinx cinerea*, in English oyster beds. (After H. A. Cole, 1942, by permission of the Council of the Marine Biological Association.)

among others a Japanese clam, *Paphia philippinarum*, which is at any rate edible and a Japanese oyster-drill, *Tritonalia japonica*, which attacks oysters both the foreign and native. The Japanese oyster was taken to Australia in 1947–8: those put down in Tasmania established safely and have bred, though it is not yet known how permanently they will be able to live there.[198] This tale could be repeated endlessly—for instance, as if the tropical seas were not already rich enough, Hawaii has had *Ostrea virginica*, *Ostrea gigas* (which both made a good start), and *Ostrea cucullata* from Australia (which died).[173] If a large corporation had been set up just to distribute about the world a selection of organisms living around or just below low-water mark on the shores of the world, it could not have been more efficient at the job, considering that the process has only been going full blast for a hundred years or less!

A good deal of chess play has also been done with clams, the often large sand-or-mud-living bivalves used for food. The Pacific Coast has now got the Eastern American soft clam *Mya arenaria* (that also lives in Europe naturally), brought by 1874, probably accidentally with oysters.[196] Hawaii has acquired two Oriental clams, *Paphia philippinarum* and *Cytherea meretrix*.[173] But these experiments are small in comparison with the great transfers of oysters everywhere. One final example of the transport of a species, but one that is not of any commercial interest, is a small Xanthid crab not more than an inch across, *Rhithropanopeus harrisi*, of Eastern North America which reached California probably with oyster materials about 1938. Here it lives in rather muddy estuarine water but only in places where occasional freshening of the water kills a native species of crab, *Hemigrapsus oregonensis*, with very similar habits. It likes to live among the calcareous tubes of the worm *Mercierella enigmatica*.[182] *R. harrisi* turned up in the harbour of Copenhagen in 1953, living with the same Serpulid worm, *Mercierella enigmatica*, also introduced there! It has reached other parts of Europe, including the Black Sea.[201–2]

In the midst of this rather complex tangle of species and dates and places we can discern the setting in of a very strong historical move, the interchange of the shore fauna of continents, and also sometimes the plankton of different seas. It is only an advance guard, yet some of the species have already taken up prominent posts in the new communities they have joined: *Biddulphia* in phytoplankton, *Elminius* in the intertidal zone, oysters at various low levels of muddy shores, their dominant enemies like *Urosalpinx*, competitors like *Crepidula*, and we should remember (from Chapter 1) the grass *Spartina townsendii.*

Some very startling explosions in marine populations have happened in the Caspian Sea. This highly modified relic of the Tethys Sea has undergone many vicissitudes before arriving at its present ecological state, yet still contains an enormous wealth of life. It is the biggest brackish lake in the world, 800 miles long, having about half the salinity of the sea and a rather different chemical composition, the lower depths sterile like the Black Sea of all except microscopic life, the northern part ice-covered in winter and inhabited by a race of Arctic seals. There is a very rich inshore bottom community and fisheries. Lake Aral, which is rather fresher, is also a marine relic. There are still in the deserts of these parts wells that have in them marine Foraminifera. Although the Black Sea is salt, its lagoons contain many of the brackish species that used to live in the Caspian Sea, and before that in the great Pontian Sea that united them all in Pliocene times. In 1934 Soviet marine biologists first suggested the deliberate introduction of animals from the Sea of Azov and Black Sea into the Caspian and Lake Aral, to help the fisheries.[203] The idea was backed by two extraordinary events of which we do not unfortunately have the complete history. At some previous time, but not very long ago, a bivalve mollusc, *Mytilaster lineatus*, from the Black Sea and a prawn, *Leander adspersus*, from the Sea of Azov got accidentally into the Caspian and multiplied colossally. These species both live also in the Mediterranean. Various fish have also been brought in, of which the grey mullet, *Mugil*, is said to have established itself successfully. But when a species of sturgeon was imported into Lake Aral it carried with it a parasite worm, *Nitzschia sturionis*, that did serious damage to another sturgeon there.

In 1937 research was being done on the physiological tolerance of a brackish water polychaete worm from the Black Sea and Sea of Azov, *Nereis succinea*,[190] and about 1940 it was introduced into the Caspian, with startling success.[160] By 1952 a whole programme of ecological work had been done on this species, because it was by then one of the dominant

23. 'The Monuments on Easter Island'. The great statues made by the earlier Polynesian inhabitants. In the left foreground is apparently the island tree, *Sophora toromiro*, now almost extinct; in the middle distance natives by their house and some cultivated plantains. (Reproduced by permission, from a painting by W. Hodges, who accompanied Captain Cook's Second Expedition, lent by the Admiralty to the National Maritime Museum, Greenwich.)

24. The grassy slopes on the outer wall of the old crater Rano Raraku on the east side of Easter Island, with some of the great statues made by the early inhabitants. (From C. Skottsberg, 1920.)

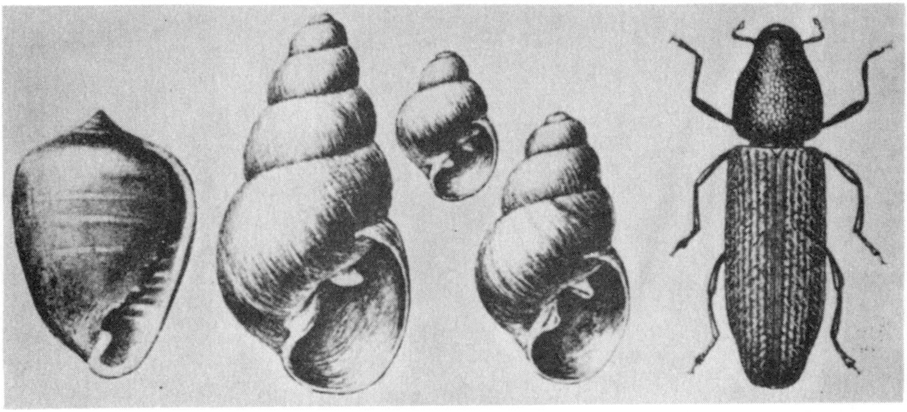

25. Easter Island has only five species of land and freshwater animals so far found to be endemic. The land snail, *Melampus pascus*, (left) and the weevil, *Pentarthron paschale*, (right) are two of these. The three land snails in the centre are forms of *Pacificella variabilis*, described as endemic, but since recognized as a Fiji species, *Tornatellinops impressa*. None of these three species measures more than 5 mm. (Snails from N. H. Odhner, 1926; weevil from C. Aurivillius, 1926; later note on *Pacificella*, see C. Skottsberg, 1956.)

26. Shells of Hawaiian land snails attacked by introduced rats. 3–7 *Achatinella*, 9–10 *Amastra*. (From J. F. G. Stokes, 1917.)

27. The Ou, *Psittacirostra psittacea*, one of the Drepaniidae, a family evolved entirely within the Hawaiian Islands. (From a coloured plate by F. W. Frohawk, in S. B. Wilson and A. H. Evans, 1890–9.)

28. Giant African snails. The large one in the hand is *Achatina achatina*, still confined to West Africa. The other is *A. fulica*, spread from its native home in East Africa by man across the Indian and Pacific Oceans. Hawaii is the furthest eastern point at which it has become permanently established. (From R. Tucker Abbott, 1949.)

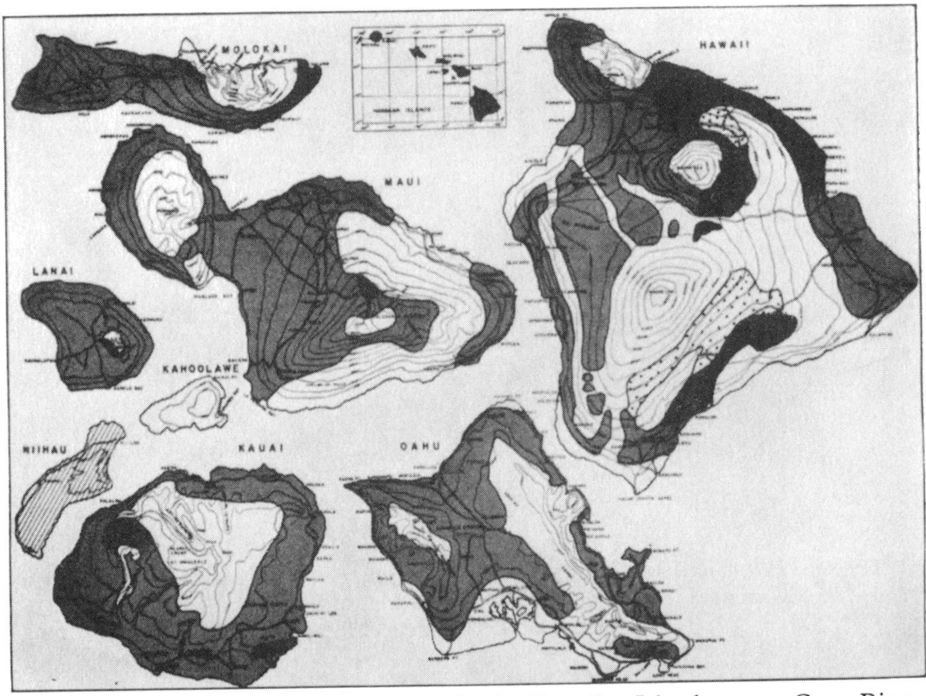

29. Distribution of introduced pheasants in the Hawaiian Islands, 1947. *Grey*: Ring-necked pheasant, *Phasianus colchicus*; *Stippled*: Japanese pheasant, *P. versicolor*; *Black*: mostly hybrids. (Niihau I. was not surveyed.) (Photographed from coloured map, with stipple added, in C. W. and E. R. Schwartz, 1949.)

30. Southern beech, *Nothofagus*, forest by the Makaroro River, Ruahine Mountains, Hawkes Bay, North Island of New Zealand. It is now inhabited by introduced European red deer, *Cervus elaphus*, and Australian opossum, *Trichosurus*. (Photo J. S. Watson.)

31. Hawaiian rats (*Rattus hawaiiensis*). The small rats of this species group have been carried, originally from Malaya, across the Pacific by Polynesian voyagers. They still survive on some islands, including Hawaii and New Zealand, but in many places have died out partly through the presence of rats brought by Europeans. (From J. F. G. Stokes, 1917.)

32. The introduced American whelk tingle or oyster drill, *Urosalpinx cinerea*, on an English oyster, *Ostrea edulis*. (From H. A. Cole, 1956B.)

33. American slipper limpets, *Crepidula*, being cleared from derelict oyster beds in England. (From H. A. Cole, 1952.)

34. Striped bass, *Roccus saxatilus*, weighing 12–17 pounds, caught by an angler in California. (From N. B. Scofield and H. C. Bryant, 1926.)

35. A striped bass, *Roccus saxatilis*, being tagged in California, for the study of its migrations. (From A. J. Calhoun, 1952.)

36. 801 slaughtered cattle being buried in a 600-foot trench, during the successful campaign against foot-and-mouth disease in California in 1924. (From C. Keane, 1926.)

37. DDT dusting by machinery on a field of potatoes in Hertfordshire during the successful eradication campaign against Colorado beetles in 1947. (Photo by courtesy of the Plant Pathology Laboratory, Ministry of Agriculture, Fisheries and Food.)

inhabitants of the benthos layer. Like *Spartina* in England, it found a zone of muddy bottom that other species had not dominated. By 1946 its populations had spread to their habitat limits in the weaker brackish waters of the Sea; and it was possible to announce that '*Nereis* accounts for a quarter to a fifth of the total calorie value of the bottom fauna of the Northern Caspian in June'. It had become an important extra fattening food for two kinds of sturgeon; it is claimed that this had come about without disturbing the balance of other benthos animals. The worms live in the superficial layer of organic material on the mud and sand bottom, where they shelter, and on which they feed. The possibilities of spread into any environment that allows play for such expansion are suggested by the fertility of a female *Nereis succinea*—eighty to a hundred thousand eggs.[190]

It remains to round off this account by giving a few instances of river-running and also of true marine fish being successfully introduced. In the North Pacific from Japan to Western America there is a group of Pacific Salmon, *Onchorhynchus*, that provide one of the biggest salmon fisheries in the world. They live in the sea but ascend rivers to breed, like our own species. There are five kinds, with various peculiar names: the chinook or quinnat, *Onchorhynchus tschawytscha*; the sockeye or red salmon, *O. nerka*; the coho or silver salmon, *O. kisutch*; the pink or hump-back, *O. gorbuscha*; and the chum or dog, *O. keta*. We are concerned with the first four species. From 1872 onwards until 1930 the United States Bureau of Fisheries, with benevolent intent, supplied over 100 million eggs of Pacific salmon to people in other countries, with the idea of establishing new salmon runs there—a considerable attempt to bring in the New World to right the Rest. The job was done very efficiently, and unlike many such campaigns, a careful record was kept of the results.[172] Many countries tried it out, though Norway refused. Because of the limited range of tolerance to water temperatures that these northern salmon have, the introductions were only successful in the northern and southern temperate zones, and failed in places like Hawaii; while in some others like the Argentine the rivers were probably too full of silt. Some sea to river runs were achieved in Chile (coho or sockeye), New Zealand (chinook), Maine (pink), New Brunswick and Ontario (chinook); while

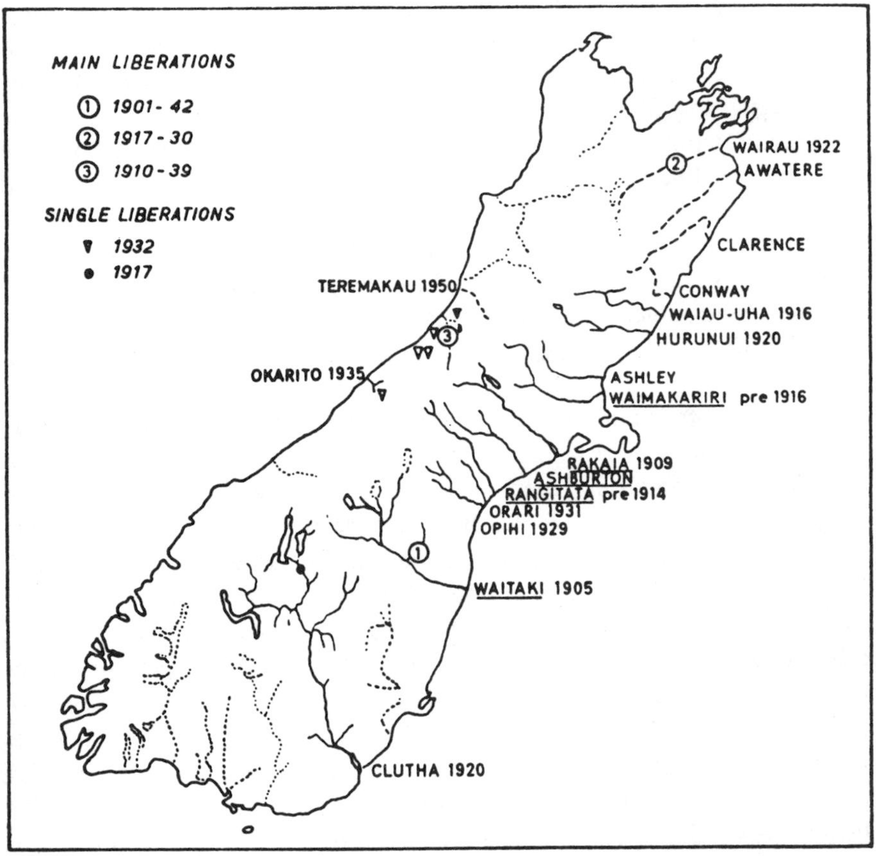

FIG. 39. Distribution of populations of the introduced Pacific quinnat salmon, *Onchorhynchus tschawytscha*, in New Zealand. *Solid lines:* well established stocks; *broken lines:* a few salmon; *dotted lines:* none. (After K. R. Allen, 1956.)

some of the populations took to an entirely inland life in lake or rivers, as in New Zealand (sockeye) and Tasmania (chinook) and certain populations in eastern North America. The quinnat (chinook) has established regular breeding stocks in New Zealand since 1905 (from eggs laid in 1901), and these occupy many rivers of the east coast (Fig. 39), ranging the seas as well, where the salmon spend a great deal of their mature life.[159] This enormous experiment has put a genus of fish formerly confined to the North Pacific into the other oceans of the world, in the belts where summer isotherms of the sea water are not above 15–20°C. After many attempts that failed in the last ninety years the Atlantic salmon, *Salmo salar*, has also achieved a breeding population in New Zealand, but only in a single river system.[159]

Between 1871 and 1880 over half a million fry of the shad, *Alosa sapidissima*, from Eastern America were planted in the Sacramento River, California, and nearly a million more in the Columbia River in 1885–6. By 1879 these fish had already begun to be abundant enough to sell and in latter years there has been an average catch every year of several million pounds.[185,196] Though the commercial fishery covers a narrower range, the shad itself now occurs from the Northern edge of Mexico right up to Alaska and Wrangell Island. Neave has remarked drily: 'Perhaps the best testimony to the fact that the shad is reacting like a native fish is to be found in recent complaints of depletion in the Columbia River, accompanied by requests for appropriate investigation of its status.'[189]

The final example of this sort of explosion of fish is the striped bass, *Roccus saxatilis*. This is a hefty fish, the official champion being one of 125 lb. from Carolina—perhaps six feet long and an angler's dream. The ordinary limit is about ten pounds, but it is apparently not rare to find them two or three times as heavy.[137] It is a sea fish but it goes into the less saline waters of estuaries to breed. Its natural home is on the Atlantic coast of North America from Florida to the Gulf of St Lawrence. In 1879 the first striped bass were brought to California, and in 1882 the only other lot, in all about 435 fish. The populations grow very fast and spread up to other places on the Pacific coast.[194] Although it is especially prized as a game-fish for anglers (Pl. 34), something like a million pounds weight of the fish were being caught in 1926, and this did not include the anglers' contribution. But since 1935 only anglers have been allowed to fish for it in California. 'The annual catch in this state since 1942 has been stable at about 1,500,000 fish. It has been estimated that $10,000,000 is spent annually on bass fishing trips and that the species provides 2,000,000 man-hours of recreation per annum.' [189] A world that begins to assess its recreation in man-hours probably cares fairly little about the breakdown of Wallace's Realms; but it will be interesting to follow the research that California is doing on this fish, to see whether its rather hesitant seasonal migrations (Pl. 35) will reach a pattern like the Atlantic one, whether the fact that it feeds a great deal on anchovies and shrimps[181] will produce effects on other fishing enterprises, how many more dominant predatory fish could be moved around in this way with success and without ill results. As Neave remarks: 'In some respects our ignorance of population dynamics is demonstrated as effectively by these successes as by the failures which have frequently attended our efforts to introduce species into new environments.' [189] It is natural to turn from the almanack of invasions in continents, islands, and seas, to a consideration of the balance between populations.

CHAPTER SIX

The Balance between Populations

In the first part of this book I have described some of the successful invaders establishing themselves in a new land or sea, as a war correspondent might write a series of dispatches recounting the quiet infiltration of commando forces, the surprise attacks, the successive waves of later reinforcements after the first spearhead fails to get a foothold, attack and counter attack, and the eventual expansion and occupation of territory from which they are unlikely to be ousted again. And it was seen that the former isolation of continents and to some extent of oceans had evolved as it were more species of plants and animals than the world is likely to be able to hold if they are all to be remingled again—almost illimitable reservoirs of species moving out to bombard other parts of the world for thousands of years to come. The impression gained might be somewhat that felt by the reader of H. G. Wells's fantasy, *The Food of the Gods*, of which he wrote: 'It spread beyond England very speedily. Soon in America, all over the continent of Europe, in Japan, in Australia, at last all over the world, the thing was working towards its appointed end. It was bigness insurgent. In spite of prejudice, in spite of law and regulation, in spite of all that obstinate conservatism that lies at the base of the formal order of mankind, the Food of the Gods, once it had been set going, pursued its subtle and invincible progress.' How one wishes that the breakdown of Wallace's Realms could have been described by Wells at the age of forty-two!

With the invasions of animals and plants that I have described, it is the successful species that are concerned. But there are enormously more invasions that never happen, or fail quite soon or even after a good many years (like the skylark in America). They meet with resistance. It is this resistance, whether by man or by nature or by man mobilizing nature in his support, that has now to be examined: what it is and how it can be understood and when necessary manipulated and increased when desired. By the end of this book I intend to carry the argument some way towards showing that we are faced with the life-and-death need not just to find out new technological means of suppressing this plant or that animal, but of rethinking and remodelling and rearranging much of the landscape of the world that has already been so much knocked about and modified by man; while at the same time preserving what we can of real wilderness containing rich natural communities. In other words we require fundamental knowledge about the balance between populations, and the kind of habitat patterns and interspersion that are likely to promote an even balance and damp down the explosive power of outbreaks and new invasions.

To study this resistance, we have therefore to look at the other side of the battlefield and see what forces are concerned. If you want to repel invaders there are three stages at which you can try to do it. You can tackle them before they get in or while they are trying, so to speak, to pass through the guard—this is *quarantine.* You can destroy their first small bridgeheads—that is *eradication.* Occasionally you may eradicate a larger population, as was done against the African malaria mosquitoes in Brazil (Chapter I), but this is a very rare event. Usually, if an invasion has got really going it can only be dealt with by keeping the numbers within bounds, that is by *control.*

Although quarantine systems are used in a great many countries to screen or attempt to screen out foreign species that may be dangerous, quite often we may not wish to keep a species out of the country at all. We don't exclude new kinds of forest trees; and it would have been a pity to keep out the large copper butterflies that were introduced from Holland into East Anglia after our own population died out. But even a forest tree may carry its own risks with it. The eucalyptus was established in California without bringing in any of the Australian insects that feed on it, because these Californian trees were grown from seed. But in South Africa and New Zealand several kinds of eucalyptus insects got in on young trees and have become pests there.[226] The United States now prohibits the importation of cherry trees from four continents because of the viruses they may carry.[224] It has been estimated also that wheat seeds can carry any of fifty-five different bacterial and fungus diseases, and some of these are not confined to wheat.[217]

I learned how easy it is to bring in a foreign insect when I carried home a few large American acorns from Wisconsin just before the War. I only wanted to have them on my desk for mementoes. A few days after I got back some chafer beetle grubs emerged from the acorns. Of course I dropped the whole lot into boiling water to kill them instantly, and that was the end of it. When the Customs officer had asked me whether I had anything to declare, it never occurred to me to say 'acorns', and I am not sure that he would have been interested if I had.

But ninety years ago a French astronomer employed at Harvard Observatory, Leopold Trouvelot, who was also studying various kinds of silkworms, brought some eggs of the European gipsy moth, *Lymantria dispar*, to his house in Massachusetts.[210] A few of the eggs or caterpillars accidentally went astray, and they started one of the major caterpillar plagues of New England (Frontispiece). They attacked and stripped the leaves off trees in forests and gardens and orchards, and in spite of immense activity and research, including the introduction of a good many parasites and some enemies, they are still a smouldering problem. By 1944 the moth had about filled the limits within which it could be abundant (Figs. 40, 41), but nevertheless heavy defoliation of trees was still a common occurrence. The foods preferred by these caterpillars include alder, apple, basswood, box elder, gray and river birch, hawthorn, all the oaks, all the poplars and willows, not to mention a score of others that are often eaten too.[204] And following all this went the deliberate introduction of a good many parasites and some enemies of the moth from Europe—quite an addition to the fauna of North America.

A friend of a friend of mine who had just returned from Egypt was rather astonished when small beetles began to hatch out of his shirt buttons. These turned out to be made from the nut of a kind of palm, and the larvae had gone on living in the stuff, having apparently passed through the manufacturing process without harm—rather like Charlie Chaplin in *Modern Times*.

So, although no one is likely to get into New Zealand again accompanied by a live red deer, we have to accept the proposition that invasions

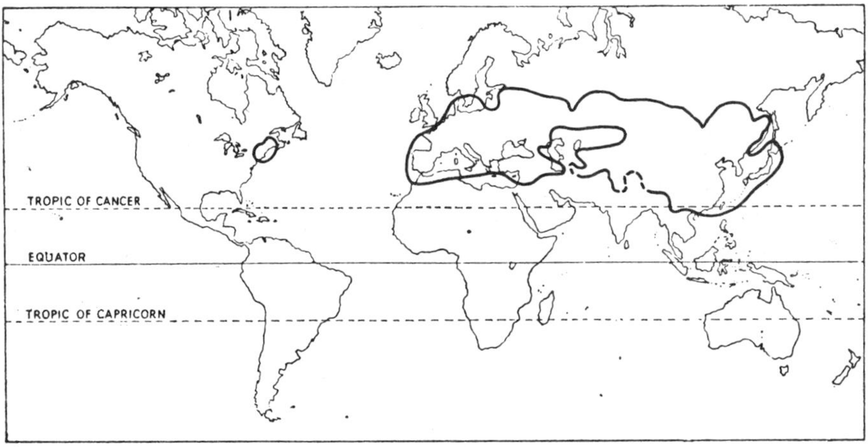

FIG. 40. Distribution of the gipsy moth, *Lymantria dispar*, a Palaearctic species that spread accidentally to eastern North America at the end of the nineteenth century. (By courtesy of the Commonwealth Institute of Entomology.)

of animals and plants and their parasites—and *our* parasites—will continue as far as the next Millennium and probably for thousands of years beyond it. Every year will see some new development in this situation. That is a way of saying that *the balance between species* is going to keep changing in every country. Quarantine and the massive campaigns of eradication are ways of buying time—though they are valuable and necessary, they are also extremely expensive. It takes so few individuals to establish a population, and such a lot of work to eradicate them later on.

There have been a score or so of very large invasions which had, from the human point of view, a satisfactory ending (and here I do not, of course, refer to the introduction of domestic plants and animals or of counterpests). They include the Mediterranean fruit-fly in Florida (1929), the African house mosquito in Brazil (1938–9, see Chapter I), the Colorado beetle in Britain (so far (Pl. 37)), and foot-and-mouth disease in the United States. The story of the conquest of foot-and-mouth disease in California was written vividly by De Kruif in his book called *Microbe Hunters*, and official annals have documented it. Up to 1940 ten invasions of this virus

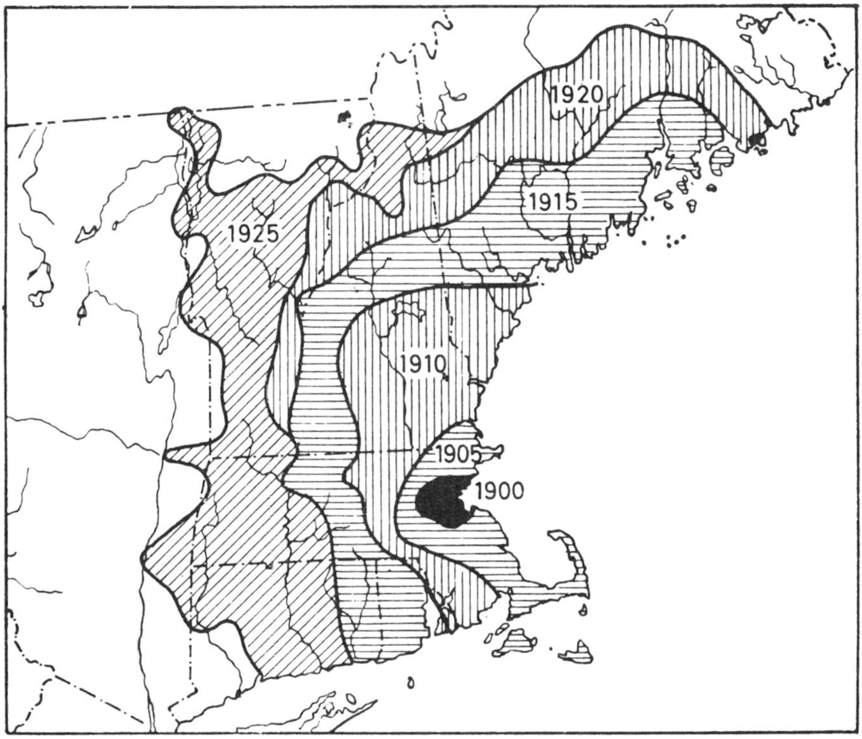

FIG. 41. Spread of the gipsy moth, *Lymantria dispar*, in the eastern United States. Its limits have extended little during the last thirty years. (From R. C. Brown and R. A. Sheals, 1944.)

into the United States had been tackled and wiped out, the greatest being one in 1914 that spread to the Chicago stockyards.[213] During that campaign 3,556 herds of cattle had to be destroyed. The various outbreaks had cosmopolitan origins, some infections coming from Asia, e.g. one in 1884 from Japan; some from South America; others from Europe. In 1924 California began to experience an invasion, whose origin is not certain, that was controlled with exceptional ruthlessness, in that not only were about 110,000 cattle and other stock slaughtered (Pl. 36), but the disease spread to wild deer in the Stanislaus National Forest, where about 22,000 were shot (about 10 per cent. having the disease showing) before the campaign closed. The total cost of all this eradication in California was about seven million dollars.[215]

In the British Isles we were able to eradicate the muskrat, *Ondatra zibethica*, before it had got really firmly entrenched in rivers and ponds (Fig. 42). This campaign succeeded in getting rid of every one within a few years, though the total numbers ever living in the country at one time did not rise above a few thousand.[231] These muskrats had escaped from fur farms, in England, Scotland, and Ireland.[230] During the Scottish trapping campaign, part of the price paid at the time for eradication was the destruction of a great many individuals of other species caught in muskrat traps set at the water's edge. Thus while 945 muskrats were killed in the systems of the rivers Forth and Earn in Scotland, 5,783 native creatures were also killed, among them 2,305 water voles and 2,178 moorhens.[220]

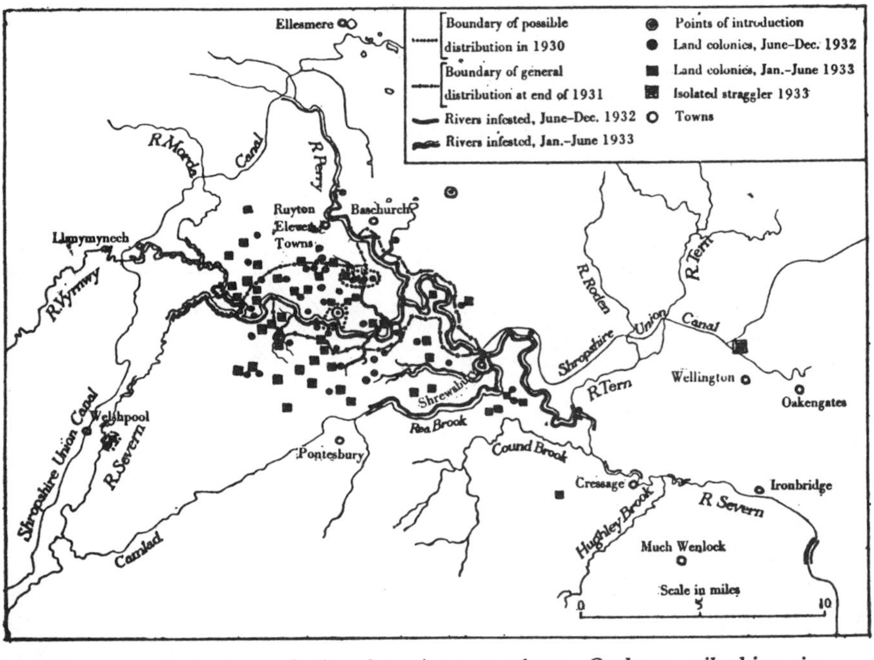

FIG. 42. Invasion of the American muskrat, *Ondatra zibethica*, in Shropshire, England, up to 1933. A year or two later the muskrat had been totally eradicated from its centres of introduction in England, Scotland, and Ireland. (From T. Warwick, 1934.)

Although ecological knowledge is of the highest value in quarantine and eradication, these operations are after all carried out chiefly by artificial methods, methods for direct killing of the organisms or changing the habitat, often powerful and ingenious, but from the ecologist's point of view not very subtle. Permanent control is in quite a different class. It means keeping numbers down to a level that prevents a species becoming any kind of dominant in the community, and damping down or even completely levelling major fluctuations and outbreaks. And it takes place not only through artificial measures but through the forces of nature. Often enough these forces alone restrict distribution and eventually produce some kind of permanent balance. No one could claim that the Canadian water weed, *Elodea canadensis*, that spread throughout Great Britain in the last century had been brought under control by man. This trailing green weed was first noticed in a pond on the Scottish Border in 1842, and arrived also in the same decade at a spot in Leicestershire, brought accidentally on American timber. Thereafter it exploded into rivers, canals, ditches, lochs, and ponds all over the country, being carried downstream, or along canals, or on birds and even occasionally (as at Cambridge) introduced as a scientific curiosity. According to Druce it reached its greatest profusion everywhere about the eighteen-sixties, but thereafter it declined considerably and universally, and has never again been considered a real plague.[208] Yet at the crest of its abundance it grew so thickly in the River Trent that fishermen could not operate their nets; at Cambridge it clogged the River Cam, interfered with rowing and made it necessary to put on extra horses to haul barges through it; one or two bathers got caught in it and were drowned; it choked a railway dock at Ely; and hindered the run-off of drainage water in the Fens.[221] It certainly could not nowadays alarm Scottish anglers[218] or render parts of the Thames impassable,[208] as was reported then. The plant is still quite easy to find living in moderate and permanent occupation of many waters. The reasons for its decline are quite unknown. They could be genetic, or indicate the exhaustion of some rare food element. In some places control was helped by cattle and waterfowl eating it.[218] But one thing is quite certain: man did not directly control this weed.

What shape of ecological world lies in front of an invading species?

If it is entering a warehouse full of stored food, there will already be a small assemblage of other animals; if into crop lands, a rather more varied community; if the crop lands still stand amongst a network of roadside meadows, hedges, and patches of wood, a very much richer system of plants and animals; if into a fairly natural woodland, an enormously complex world. Most people simply do not know the astonishing richness in species and the huge numbers of individual animals living together in one place. This is partly because they are to a great extent hidden under cover—the evolutionary result of an intense conflict between enemies or parasites and their prey, driving animals under cover, combined with the physical advantages that cover gives in other protection. It is partly also because most of the species are rather small creatures like insects and spiders and mites, or—in water—insects, crustacea, rotifers, and Protozoa. The fungal microflora is more varied than the world of larger green plants, though it does not of course provide the major structures of vegetation that set the visible scene for most communities. And in water, microscopic algae are more varied and numerous than flowering plants.

For over twelve years I have studied a small hilly area close to Oxford and kept records, a great many of which come from other ecologists in the University who have been working there in ecology or other sciences.[209] Wytham Woods cover less than two square miles. The country is quite an ordinary representative bit of English Midlands: woodland and fields, streams and marshes, a few patches of limestone grass on the top, and the River Thames flowing round two sides of it. We already know that in this ordinary (and therefore quite beautiful) bit of English countryside, only moderately spoiled so far by the progress of twentieth-century forestry and agriculture, something like 2,500 species of animals exist, and that there must be many more than this still to be observed. The number of individual animals on the whole area undoubtedly runs into thousands of millions, because we know a little about this from population counts that ecologists have already done there.

Naturally, most animals colonize one or only a few habitats, not the whole lot; but even so they will find themselves entering a highly complex community of different populations, they will search for breeding sites and find them occupied, for food that other species are already eating, for cover that other animals are sheltering in, and they will bump into them and be bumped into—and often be bumped off. Besides this, each habitat shares part of its fauna with neighbouring ones. An ecological system, like an organized human community, has its separate centres of action—such as the soil and the tree canopy, the marsh and the stream, the fallen log and the bird's nest—but always at some point you can find connexions between them, and these may affect the balance between populations. The invader is therefore working his way somehow into a complex system, rather as an immigrant might try to find a job and a house and start a family in a new country or big city. The shortest way of describing this situation (and a convenient one, provided we remember that it largely describes ignorance and not knowledge) is to say that it is meeting *ecological resistance*. The question is, what is this? And why is it suddenly overcome by certain species?

This resistance to newcomers can be observed in established kinds of vegetation, indeed competition is one of the central concerns of plant ecologists, competition for light and soil chemicals and space; though by arrangements that even advanced plant ecology has not yet revealed very far, fifty or more different species of plants may be found living permanently together in one type of vegetation. But by far the greater part of our alien plants live in habitats drastically simplified by man, including of course our crop plants: in arable land, waste dumps, railway tracks, walls, and so forth. 'Except cornfield weeds, few introduced species have really established themselves sufficiently to form part of the British flora' [223]—that is to say, they have not penetrated the natural closed vegetation of Britain. It is now known that some of our weeds that flourish or used to flourish on the open ground of cornfields and roadside waste, were commonly distributed in this country during the early part of the Post-glacial period.[212] When they first recolonized Britain it was on to the open tundra ground where plant competition was not so high as it is in a fully developed meadow or wood. Then they decreased a great deal. And at least two species of snail, *Succinea oblonga* and *Catinella arenaria*, underwent the same experience. Formerly their range in Great Britain, as shown by fossil shells, extended inland. But they are now confined almost entirely to coastal habitats, e.g. *C. arenaria* on the sand dunes at Braunton Burrows in

Devonshire. The sea plantain, *Plantago maritima*, also had a history rather like this, and now lives by the sea and also on a few mountains. The scentless mayweed, *Matricaria inodora*, was a Late-glacial plant that is now a weed of fields and waste land. It has a closely related subspecies that is a common maritime plant.[212]

The white dead-nettle, *Lamium album*, is one of the plants that has never been admitted into the natural vegetation of Britain. This labiate resembles a stinging nettle but has no stings, though its leaves have an excessively acrid taste. It has white flowers that are visited through much of the year by bumble-bees, and are an important element in their early spring diet. The plant grows abundantly on roadsides, the edges of arable fields and in waste ground, but always outside our native communities: it is scarcely mentioned by Tansley in his *Vegetation of the British Islands*. But in its original home in the Caucasus the white dead-nettle is a successful woodland plant in a rich community of other species. Yet it has doubtless been in England for hundreds of years without making the grade here as a woodland plant. But supposing it had done so, would it have replaced some native species, or just added one more to the list? This is a recurrent and an insistent question that keeps rising in the mind, and perhaps is the single most important problem lying underneath all the facts of the present book.

Introduced animals often do replace or reduce the numbers of native ones. This is seen in its simplest form in the sea, where many animals compete for space on the shore and the sea bottom, collecting plankton or organic detritus from the ocean water. The oyster beds referred to in the last chapter illustrate this, but it is not only oysters with which a foreign invader comes in contact there. This can be realized from the ecological survey of two oyster beds done quite recently in Essex estuaries by Mistakidis—the first thorough investigation of its kind on a community that has been suffering ecological troubles for over fifty years.[219] He took two patches, one of 15 and the other of 33 acres, and found that the lists of animal species ran to 92 and 113 respectively, though most of these were only scattered in small numbers or colonies. But in each place 24 species occurred at more than half the small sampling stations studied. This, therefore, is the oyster bed community—the system for which the German ecologist Möbius many years ago first coined the descriptive term

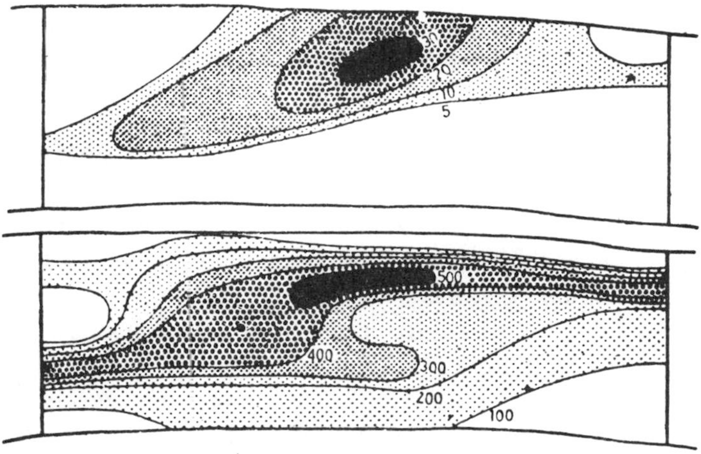

FIG. 43. A population map of the English oyster, *Ostrea edulis* (*above*), and its accidentally introduced competitor the American slipper limpet, *Crepidula* (*below*), on a thirty-three acre oyster bed in the River Roach, Essex. This bed had been neglected, and only young oyster spat were represented. The figures are the population numbers per square metre, the average numbers of *crepidula* being 179. (After M. N. Mistakidis, 1951.)

biocoenosis—into which *Crepidula* comes as a competitor. In these Essex beds the slipper limpet was the dominant animal, having already reached the remarkable average population densities of 446 animals to a square metre on the first and 179 on the second patch (Fig. 43)—equal to live weights of ten and four tons per acre. This limpet builds up a series of animals, living one on top of the next, in these particular populations from two to eight, but running exceptionally to eighteen in number. It is curious that both human beings and this mollusc on the eastern seaboard of North America have evolved the same skyscraper principle for exploiting valuable ground to the full. For the slipper limpets it means that more food is processed and the rate of reproduction high. It should here be explained that this Essex survey was done on neglected oyster beds, that had been more or less uncultivated for at least eight years, and that only a little restocking with oyster spat had taken place. But, as can be seen, a neglected oyster bed is quite rich in other species, though many in turn will no doubt be affected by the further spread of *Crepidula*. In well-managed beds the slipper limpets have to be dredged up and smashed by machinery—in fact weeded out (Pl. 33). 'The ground will revert quickly to the condition aptly described as "mud and limpets" if dredging and removal of *Crepidula* is suspended and will again reach its climax in from ten to fifteen years. On well stocked oyster beds the bottom community in Essex may be described as characterized by *Ostrea* with *Elminius* as epifauna, since this barnacle is now ubiquitous on both oysters and cultch and apparently has not yet reached its maximum density.' [219]

Competition for space is also common among more mobile animals than these. Although the European starling seems to have found a rather new feeding niche for itself in the United States, that does not bring it into very active competition with native birds, yet it has driven away some of them from the limited breeding sites that exist in towns. As a result the

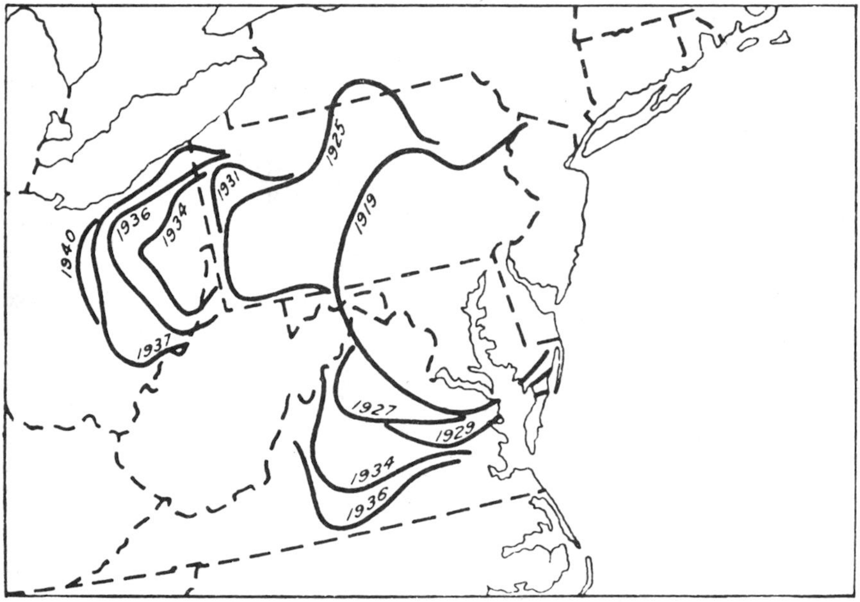

FIG. 44. Invasion of the eastern United States by the black grain stem sawfly of wheat, *Cephus tabidus*, which arrived from Europe by 1889. (From E. J. Udine, 1941.)

blue-bird, *Sialia sialis*, and the flicker, *Colaptes auratus*, are now less common where the starling occupies the towns, though their populations as a whole are safe enough.[17] The two species of wheat stem sawflies that have also colonized the United States have shown a very interesting progression.[229] The European wheat stem sawfly, *Cephus pygmaeus*, had arrived by 1887 and afterwards spread over some of the north-eastern states. The black grain stem sawfly, *Cephus tabidus*, also from Europe, was in New Jersey by 1889 and its spread westwards and southwards was steady and had in 1940

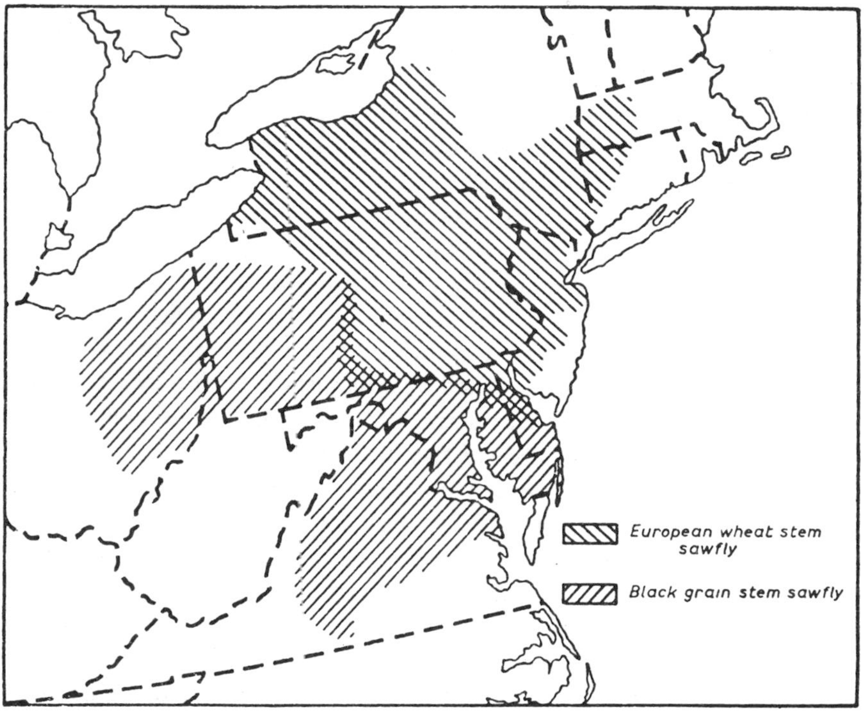

FIG. 45. Mutually exclusive distribution of two introduced European stem sawflies, *Cephus pygmaeus* and *C. tabidus*, in the eastern United States. The latter originally spread farther north into the range already occupied by the former. The narrow zone of overlap may be determined by competition of the sawfly larvae for space in the wheat stem. (From E. J. Udine, 1941.)

reached the limits shown on the map in Fig. 44. During its expansion the second species invaded part of the territory already held by the first one north of it, but in Eastern Pennsylvania the European wheat stem sawfly was gradually replacing the newcomer, and the other map (Fig. 45) gives the result, with two separate ranges just overlapping in a fairly narrow zone. 'Recent observations indicate that in areas where both species are present *Cephus pygmaeus* adults emerge about a week earlier than those of *C. tabidus*. In view of these observations, and since more than one sawfly egg is often found in a wheat stem, although only one larva reaches maturity, it may be assumed that the larvae of *C. pygmaeus* destroy eggs and ensuing larvae regardless of species. This relationship may account for the reduction of *C. tabidus* in the areas where *C. pygmaeus* is present.' Some of the classical experiments done on mixing together laboratory populations of two different kinds of insects that have larvae inside wheat grains have given just this kind of result.

I have set out these examples where competition is for space to live and eat or breed, because they are the simplest ones to understand. But it is likely that competition is usually a far more complicated matter. When we talk of 'competition' a careful distinction must be underlined. In the ordinary colloquial sense, it means a direct struggle between the individuals of two species. This is usually called by ecologists 'interference'. But there may be many indirect influences, acting through other species like parasites or enemies, or the relative skill of two kinds of animal, or a whole string of causes and effects that can be very hard to trace. These, equally with interference (if they decisively affect breeding or survival) may lead to the replacement of one species, or part of the populations of one species by another—a demographic event of whose interior causes we may be and usually are almost ignorant. The snag about handling these perfectly genuine concepts is that replacement may occur without direct interference —as suggested above, or when one species decreases through pure coincidence from independent causes during the increase of the other; and interference may occur without replacement, as when two species jostle for nesting sites when there are plenty more around for the loser to occupy. When there is a very clear verdict, as with the Argentine ant, that hard fighting has resulted in regular and catastrophic and general replacement of other species, including other ants, the case is complete: that is replacement through interference. With the white dead-nettle we see its failure to penetrate highly organized close vegetation, but we do not know why. With the English oysters we see them defeated by a combination of circumstances, amongst which competition for space with an invading American mollusc and an invading Australian barnacle can be seen by direct observation to be very important. But it must be remembered that there are many other forces operating here, such as the invasion of their new enemy *Urosalpinx*, the differential effect of cold winters, and also other mortalities like that among oysters in 1920, of whose causes we know practically nothing.[191] Our ignorance of the nature of competition is also illustrated by the history of the red and grey squirrels in England. The American grey squirrel, *Sciurus carolinensis*, has replaced our native red one, *Sciurus vulgaris*, in the Midlands and part of the South of England.[225] This sometimes happened quite quickly. But often it has taken up to fifteen years for the process, whatever it is, to be complete. For we still have no notion what happens, except that there is a change. It is therefore natural to be cautious; and yet there is the great blank area where no red squirrels have returned, and this is where the grey ones first spread and are now permanent inhabitants. Outside it there are plenty of red squirrel populations still, though they have fluctuated, often severely.

Sometimes foreign species have been able to edge in without producing any noticeable disturbances or making our own similar species extinct. Several kinds of fresh-water shrimps have been quietly spreading in our rivers and canals during the last twenty-five years. *Eucrangonyx gracilis* comes from North America. I have found it transported locally from one pond to another at Oxford on water plants for an ornamental pool, and this is probably how they came from America. This shrimp is rather slender, bluish in colour, and usually walks on its front, whereas our common British *Gammarus pulex* is stouter, brown, and often lies and swims on its side, especially when the male and female are coupled. It has spread widely in England and Wales.[228] Another North American shrimp, *Gammarus fasciatus*, is very locally established in England and only in saline water, though it comes in fresh water in Ireland. Two others come from the Caspian region originally. *Corophium curvispinum* is really a saline and brackish water shrimp of the Caspian and Black Seas, that has in latter years colonized the fresh-water rivers of Europe. The farthest western point of its dispersal in 1935 was at Tewkesbury, Gloucestershire, where it was living in mud tubes in the River Avon.[207] The other is *Orchestia bottae*, a lively shrimp that can jump. This has been found in the River Thames,[205] in a river in Norfolk,[222] and in Yorkshire.[211] Here are four new additions to the British shrimp fauna, three of them able to colonize fresh water. We have as yet only vague indications about what their fate will be and how far they will react upon our own species. But it seems pretty certain that competition can occur among shrimps.[214] Our *Gammarus pulex* is absent from Ireland and some of the small islands of the west coast of Britain. In these places another species, *Gammarus duebeni*, occupies the ecological niche of *pulex* in fresh water. But over nearly all the mainland of Great Britain it lives only in brackish and estuarine waters. In the Lizard Peninsula of Cornwall *duebeni* occupies fresh water, and *pulex* seems to be absent. The only place where both occur living together is the Isle of Man, where possibly the balance is in the process of change. Therefore, when new species arrive and spread, even if they do not have the appearance of the explosive invader, they may herald the onset of future changes in the balance of populations. The complete unravelling of any of these relationships will be an interesting but often very difficult task.

CHAPTER SEVEN

New Food-chains for Old

The natural living world is arranged in very complex channels of supply that are known as food-chains. From the plant through different species of animals there are usually several, often as many as five stages, but seldom more than that. Alfred Lotka called these chains of species connected energy transformers, because each species was using up in maintenance, movement, and increase some of the energy originally captured by plants from sunlight, and passing it on to another in the cycle of supply. 'The entire body of all these species of organisms, together with certain inorganic structures, constitute one great world-wide transformer. It is well to accustom the mind to think of this as one vast unit, one great empire.'[257] Each species degrades the organic energy into heat, or else its body is devoured alive or dead. Even the animals right at the end of the food-chain are devoured when they die. The living plant is usually able to keep the greater part of itself intact while it is alive, although a not inconsiderable fraction of it passes into animal food-chains. But probably much the greater volume is handed on after the plant dies or in the leaves it sheds, and to a lesser extent when the animal dies. Of course, if this were not so, we should not behold the solid mass of green vegetation, the living basis of all communities would be weak, and the life of whole communities very precarious, which certainly is not generally so in natural ones. While they are alive plants and animals may shed part of their bodies (as with leaf litter, pollen, or moulted insect skins), give off secretions (as with nectar and aphid secretions) or excretions (especially impressive with large ruminant animals, though just as important though less obvious in others). These all create further loop channels in the ecosystem. The immensity and complexity of all these channels, or connected energy transformers, can be imagined, but is very far from being understood except in outline.

It is essential for us to know what role they have in the regulation of population size and density, because nowadays they are perpetually being altered and damaged and new species substituted for others.

The first person to draw a picture of food-chains was Peter Brueghel the Elder. He had an exact artist's mind, his paintings are full of scenery and action and people's occupations and of colour, and at times they reached a rather nightmarish insight into nature, and the nature of man. This astonishing drawing (Fig. 46), turned into an engraving, was done in 1556.[249] It includes what might well be an early ecologist carrying out a food analysis (apparently with a very large bread knife), and what might be a very early applied biologist hurrying away to the north-east of the picture, having partially turned into a fish by absorbing fish that have also eaten fish (which is true in a strictly limited sense). Altogether this picture tells one more and makes one feel more about the supply lines of nature

FIG. 46. 'The big fish eat the small ones'. (Engraving from a drawing by Peter Brueghel the Elder, 1556. From G. Gluck, 1936.)

than any amount of formal logic might do. In medieval times there seem to have been proverbs used that were the origin of Brueghel's drawing. There is cross-talk between two fishermen in Shakespeare's *Pericles*, about the big fish eating the smaller fish. As I have mentioned in an earlier book, the Chinese people had shrewd ecological proverbs about these matters.[245] Worthington found also that the Banyoro natives of Lake Albert catch their fish with great skill by using a successive food-chain of baits.[274]

But with land in cultivation, whether pastoral, ploughed, or gardened, the earnest desire of man has been to shorten food-chains, reduce their number, and substitute new ones for old. We want plants without other herbivorous animals than ourselves eating them. Or herbivorous animals without other carnivorous animals sharing them. Only in the sea do we still depend on nearly full natural food-chains to supply our wants: the plaice eating the bivalve mollusc that feeds on debris, the herring that catches plankton in an intricate community of other species, and the whale that eats euphausid crustacea that depend on smaller plankton food. The three propositions given above seem so extremely simple at first sight, and have after all provided food and materials for a vast human population. Clear the jungle or plough the prairie or cultivate your oyster grounds. Keep down or kill or drive away all competitors. Shorten the food-chain and harvest more energy. Improve the domestic or semi-feral stock. We do not always shorten the food-chain completely, by being vegetarian, because our poor digestion, our tastes, and the concentration of certain chemical virtues are somewhat in conflict with the most economic method of harvesting just calories. I have watched wood-ants in a grove of birch saplings in the New Forest keeping small flocks of plant-lice on the twigs. The ants had killed absolutely every other insect on the trees. These were their pastures, and sentinels stood by each flock of aphids, while other ants came and milked them for sweet excretions. I broke off some small twigs, and the sap began to flow out, and quite soon some of the ants left their herds and collected the sap at this direct source. No doubt we should often eat grass and dispense with sheep and cattle, if our digestions would permit.

Some of the profoundest changes in food-chains have come about through the introduction and spread of domestic grazing animals. A hundred years ago the grass plains of North America (Pl. 38) were still occupied by huge roaming herds of bison. 'Buffalo Bill' only died in 1917. The bison was the chief grazing animal in the centre of the continent, but in a comparatively few years was completely replaced by cattle and sheep, as well as by other kinds of farming. The structure and composition of the prairie vegetation also changed. But a species of bird, *Molothrus ater*, that used to accompany the buffalo apparently in order to catch insects disturbed by their trampling, and frequently rode on the beasts, transferred its attachment. 'The Buffalo Bird of the plains of the pioneer days is the Cowbird of the farm pastures of today.' [247] The old photograph (Pl. 39) of thousands of buffalo skulls stacked up gives a vivid notion of the scale of this ecological replacement. Here the replacement of the bison by domestic stock happened indirectly, through the choice of food by man as a predator, combined with the symbiosis he keeps with his domestic animals. This whole transaction was on the grand, the continental scale, millions of bison being replaced by millions of sheep and cattle.[261] Having got this new set of food-chains, it might seem quite simple to maintain them in equilibrium. They are short and easily understood. Yet in many parts of the world it is just this equilibrium that has broken down through overgrazing or mismanaged grazing, and soil erosion has often completed a process that may end with the shortest food-chain of all—nothing: what mathematicians like to call 'the limiting case'. Such a condition, even in a region of Canada very favourable to good farming, may be seen in the curious photographs in Pls. 40 and 41. And if people cannot manage this very straightforward chain of linked populations, it is not surprising that more complex ones give trouble.

According to William Vogt, writing in 1948 about the American plains: 'The western range lands, comprising nearly 800,000,000 acres, support almost 75 per cent. of the nation's sheep and more than 50 per cent. of its cattle. Originally the grazing capacity of western lands was able to carry about 25,000,000 head but the vegetation has been so seriously damaged by overgrazing that by 1935 the capacity had fallen by half. Since then, largely because of an increase in precipitation, the range has made a partial comeback . . . No less than 589,000,000 acres are eroding—more or less seriously.' [267] Starker Leopold and Fraser Darling have traced the extra-

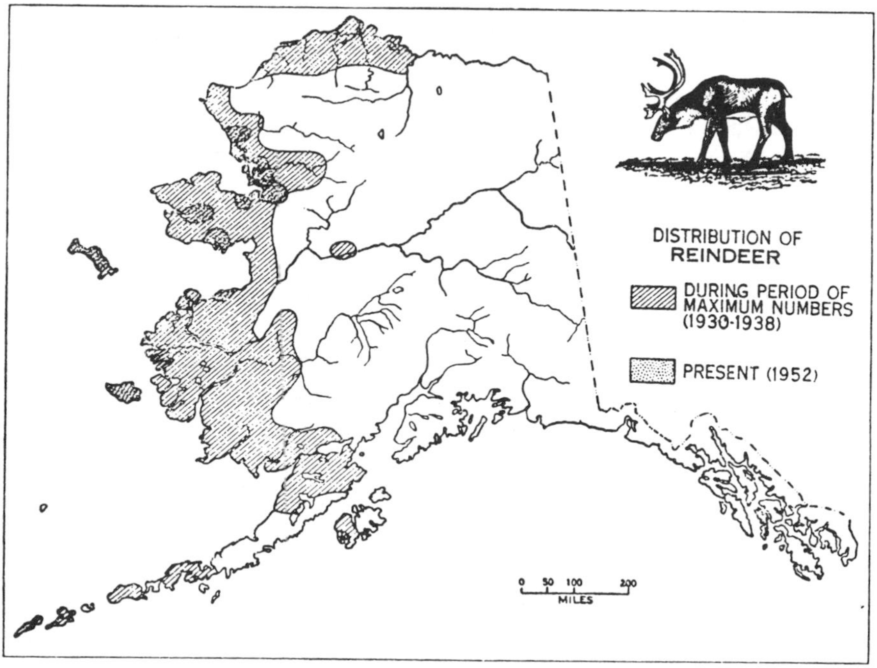

FIG. 47. The reindeer was brought to Alaska at the end of last century, and increased to over half a million, after which the numbers declined catastrophically, mainly through the overgrazing of their winter lichen food. (From A. S. Leopold and F. Fraser Darling, 1953.)

ordinary history of reindeer pasturing in Alaska. These animals were brought from Lapland in 1891–1902 to make a new resource for the Eskimos, and they increased and spread to something over half a million animals.[256] At the present time there is not more than a twentieth of that number left (Fig. 47). The reason seems plain: they were allowed to eat off the lichen supplies that are essential for winter survival; lichen grows very slowly, complete recovery needing at least twenty-five years, and its ecology is quite complicated, for other things like fire have also played a big part.[260]

The use of living species to capture energy and make special substances for us is still the central industry in the world, because we still cannot make synthetic food at any reasonable cost. Scientists have already bypassed some other natural channels of supply, for example vegetable dyes, silk, horsepower, and carrier pigeons. Some of the new substances or machines have largely replaced the original species agencies, though the horse-power has to be fed with oil. The geological banks of oil and coal have made the manufacture or running of most of these substitutes possible; and as long as there is something in the bank account this is much simpler because it avoids the complication of having to manage species. There is still an organized world trade in tannin for making leather. This needs the natural tannin that happens to be concentrated sufficiently in certain plants (herbs, shrubs, or trees), and these come from every continent to make a world market in the stuff. But tannin for some purposes can now be made synthetically. This in itself will alter the food-chains in five continents, where the tannin-producing plants are collected or farmed. Nevertheless, the enormous problem still is to manage, control, and where necessary alter the pattern of food-chains in the world, without upsetting the balance of their populations. It is this last problem that has not by any means been solved, and which is exacerbated every year by the spread of species to new lands.

In the Neolithic days of animal ecology, that is to say about twenty-five years ago, it seemed reasonable to suppose that every natural food-chain contained within itself the explanation of the control of populations. E preyed on D, D preyed on C, C preyed on B, B was a herbivore that ate the plant A. Each higher consumer layer kept down the numbers of the one below, and each one below limited the numbers of the one above through food supply. That this argument does not go quite in a circle was pointed out independently about this time by two mathematicians, Lotka[258] and Volterra,[268] whose equations and suppositions made a deep impression on their contemporaries. Being mathematicians, they did not attempt to contemplate a whole food-chain with all the complications of five stages. They took two: a predator and its prey. The arguments then went on to show that, in effect, each took turns to control the population of the other, with resulting fluctuations in numbers. Because this theory came at a time when the occurrence of such fluctuations had already been noticed in nature it seemed reasonable enough, though this really supplied no firm proof. But thirty years later we have more facts to test it with, and there does not seem much doubt that theories that use the food-chain for an explanation of the regulation of numbers are oversimplified, and often just untrue for certain species. There are other forces at work, not omitting chance disasters, and—perhaps more commonly than we have formerly believed—various methods of regulation operating through the population of the species itself.

Nevertheless the potential power of food-chains is undoubtedly unleashed in many instances where counterpests are used for bringing about control of populations. That is, when an invading species reaches too high a level of abundance, it can sometimes be reduced by the introduction of a predator (Pls. 42, 44) or parasite (Pl. 43), or for weeds a herbivore. This highly technical field of activity has reached very wide proportions and there is now a continual traffic of introduced counterpests to every country of the world that has any crop-growing or forestry. Some instances have already been mentioned. With the fluted scale insect and the prickly pear the operations were startlingly successful; with the European spruce sawfly highly promising; with the Japanese beetle and the gipsy moth incompletely so. I have cited examples also from New Zealand and Hawaii to show how species may be brought from any part of the world for this purpose, provided there is the faintest likelihood that they will work. All this is changing the species networks of the world.

Many counterpests fail to establish at all. Many more become residents of the new country without necessarily producing control of the pest. The proof that the apparently successful ones have done the job without assistance from unknown causes and events is usually pretty rough or even lacking altogether. But there can nevertheless be no doubt that counterpests have done splendid work in ameliorating disastrous situations. In such work the biologist is not shortening but lengthening the food-chains, but he takes care if possible to lengthen them by only a single extra link. For example, the outstandingly successful conquest of the prickly pear problem in Australia, chiefly by means of a moth, *Cactoblastis cactorum*, introduced from South America, might have failed had any of its parasites come in with it or had the native Australian insect parasites been able to kill more of the caterpillars than they do—less than 25 per cent.[242] But there are still invading plants like ragwort in New Zealand and St Johnswort in

Australia that have not been controlled by introduced insects, though the insects themselves have got established.

Sometimes an invasive plant may gradually be brought under control by accidentally introduced animals following it later on. We may be seeing the beginning of this in *Rhododendron ponticum*, the large purplish-flowered species from the Mediterranean region, brought to Britain by 1763 and now well established as a natural shrub on sandy and acid soils and even as underscrub in woods. For example it is replacing the holly extensively in Killarney oak woods.[269] Foresters now look on it as a serious weed that hinders the regeneration of trees.[237] It is seldom attacked by rabbits, which has given it something of the competitive advantage over other shrubs shared by our elder and hawthorn. There is an unconscious extra truth in the words of a bee expert who wrote: 'Where an informal or unpruned screen is desired and there is ample space the common Pontic Rhododendron (*R. ponticum*) is hard to beat.' [253] Here he is speaking of their use as windbreaks. The honey of this species is sometimes poisonous, as Xenophon long ago recorded during the journey of his army in Asia Minor. But in England there have been hardly any cases of poisoning, probably because hive bees cannot easily reach the nectar in these flowers, though ong-tongued bumble-bees visit them regularly.

What happens to a shrub brought from abroad 200 years ago, and with these invasive qualities? A few native insects have influenced its growth, though not in any decisive way.[272] When the rhododendron grows in woods, it may be attacked by two kinds of weevil, *Otiorhynchus singularis* and *sulcatus*, that are well-known damagers of fruit bushes, and by certain moths of the family Tortricidae that includes some severe oak defoliators. Some of these moths eat the leaves and one curls them up to pupate in after having developed on the oaks above. A few other native moths and beetles have been found, but none on any large scale. Besides these, four foreign insects have established themselves on rhododendrons in Britain.[271, 272] The natural distribution of the whole genus (with which is more or less incorporated *Azalea*) is in North America, locally in Europe, and again in Asia and extending as far as New Guinea and one species in North Australia (Fig. 48). There is therefore a very large reservoir of possible insect immigrants, taking into account that there are more than

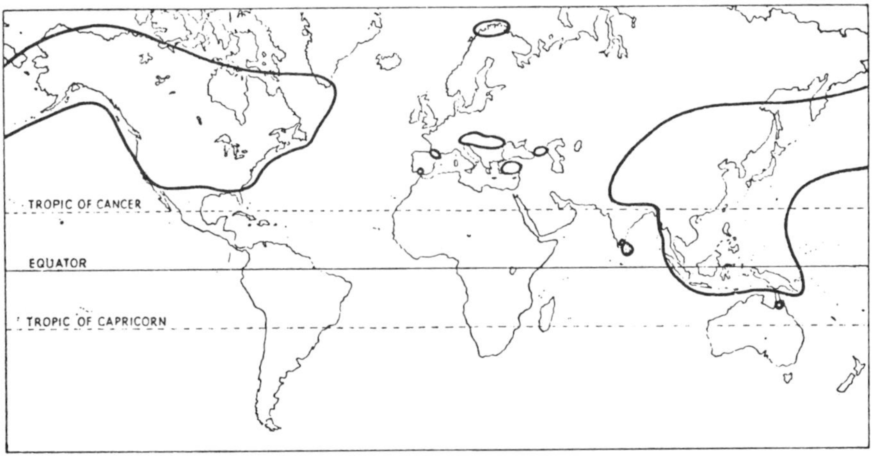

FIG. 48. Natural distribution of the genus *Rhododendron*, whose head-quarters are in south-east Asia. (After G. Fox-Wilson, 1939A.)

900 species of the host-plants known. There is an Oriental moth, *Gracilaria azaleella*, native in Japan and now spread to North America and Europe, and recorded for England. An Aleyrodid 'fly', *Dialeurodes chittendeni*, probably from the Himalaya, though this is not certain. It has spread into the United States and Canada, and in the nineteen-thirties came from America to England, where it has spread from the south and south-east. The honey-dew that its larvae excrete makes a culture-medium for the growth of sooty moulds on the leaves. The third arrival is a Tingid bug, *Stephanitis rhododendri*, first known here in 1901: it has spread very widely (Fig. 49) and attacked many species of *Rhododendron*, including *ponticum*, though this is not one of its favourite foods. This bug sucks the under sides of leaves, deranging the tissues and making them mottled, a condition known as 'rust'.[273] It is a North American insect, that reached Europe before it came here. The other and most important species is the large brilliant red and green Jassid bug, *Graphocephala coccinea*, also North American, which reached England by 1933 and has colonized the south and south-east counties.[233] These bugs chiefly suck the top surface of the leaves, usually through the veins: they seldom feed on flower buds. But they lay their eggs in a scar rasped by the female in the lower scale of

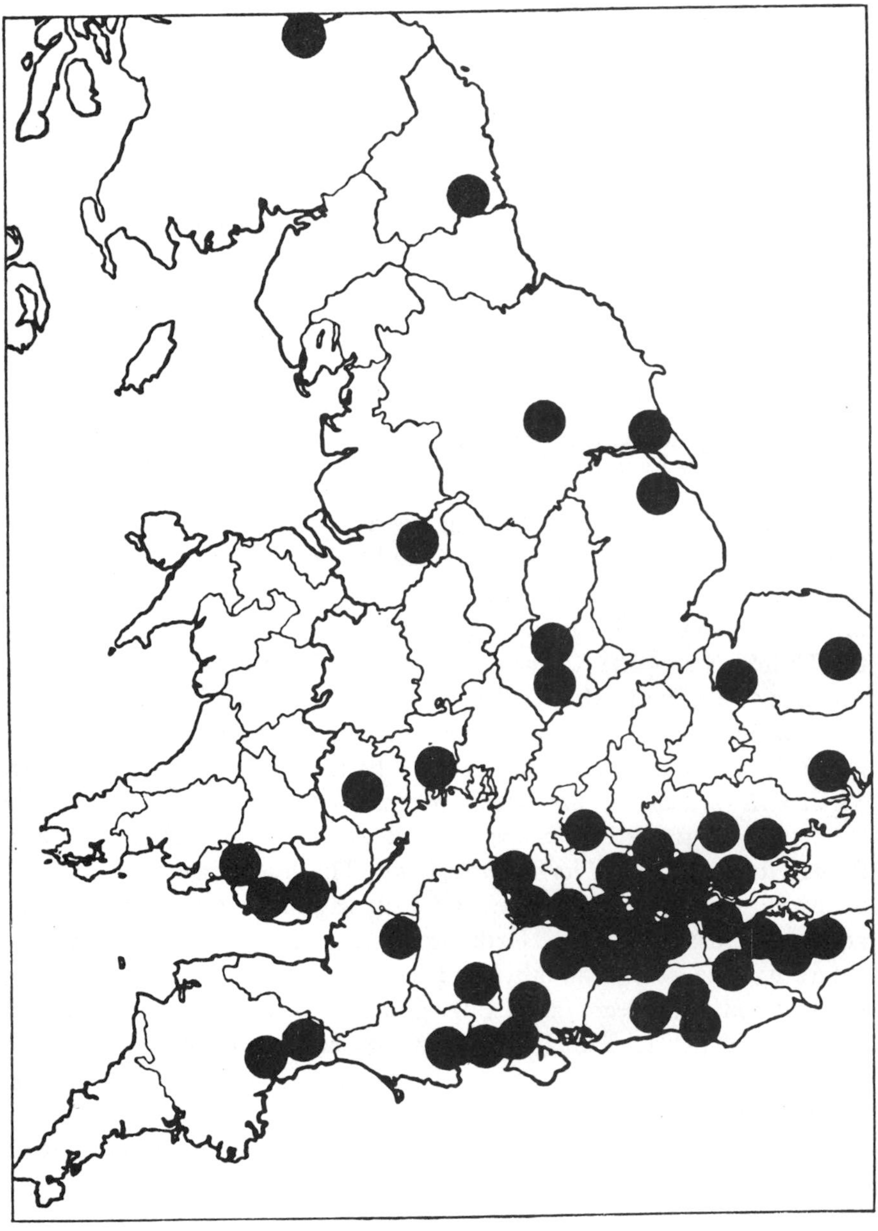

FIG. 49. Distribution in Britain of a Tingid bug, *Stephanitis rhododendri*, accidentally introduced from North America via Europe. It feeds on the leaves of *Rhododendron ponticum* and other species, and damages them. (From G. Fox-Wilson, 1939B.)

these buds, from which the young hatch in the spring. Coincidental with the spread of *Graphocephala* in England has been a disease known as 'bud blast' also apparently from America,[243] that starts in the autumn and kills the bud entirely through infection by a fungus, *Pycnosteanus azaleae*. It is thought to be extremely likely that the disease is inoculated by the egg-laying of the bug, though direct proof has still to be given.[236]

In all these events, haphazard in origin, developing at different times and different speeds, we can surely see the building up round this plant of a natural community of herbivorous occupants, one carrying a fungal disease and all reducing vitality through sucking the juices of the leaves or eating them. They come from Asia and America, and are joined by a few British residents. Is *Rhododendron ponticum* invasive partly because it has not brought its food-chain with it? And how soon will these insects acquire parasites and enemies or begin to compete amongst themselves? Is this already happening perhaps? Compare this incipient gathering on *Rhododendron* leaves with the picture worked out by Tilden for a native shrub of sand dunes in California and Oregon, *Baccharis pilularis*.[266] Associated with it he found, by two years of systematic observation, 257 species of arthropods of which 221 were insects. Of the latter 65 were parasites and that was not complete. There were 53 species of primary herbivores—leaf-nibblers, leaf-miners, stem-borers, leaf-suckers, root-feeders, gall-makers; 23 species of predators; 55 species of primary parasites, 9 of secondary parasites, and even one tertiary parasite. *Rhododendron ponticum* still has far to go in acquiring a fauna here, and of course it may not survive this increasing barrage of natural selection.

This rhododendron has seized an ecological position in Britain in competition with other shrubs, the Argentine ant in most continents in competition with other kinds of ants. But sometimes a species arrives and finds an ecological niche not occupied at all by any similar form. This has happened in some of the mountain forests of Cyprus.[270] The ship rat, *Rattus rattus*, must have reached the island hundreds of years ago—it is supposed to have arrived in Western Europe by the early Middle Ages. In Cyprus it does live in human settlements, but also far away from them in the macchia and forests. Where the latter consist of Aleppo pine, *Pinus halepensis*, the rats eat its pine-cones, a habit quite unusual for this species, though common in squirrels. But in Cyprus there are no squirrels. The ship rat originally comes from tropical forest in Asia, and in many parts of the world lives not in trees but in the roofs of native houses; and even in Britain it lives much higher up in buildings than the Norway rat. In the course of its thorough exploration of Cyprus it has found a vacant niche. Crossbills, *Loxia curvirostra*, do eat pine-cones in Cyprus, but of another species, *Pinus nigra*. This rat also lives up palm trees on many Pacific islands, and eats young coco-nuts.

Some biological control consists of trying to break a closed chain of symbiosis between an ant and an aphid or scale insect. An experiment was done in California upon the Argentine ant in citrus orchards, where it is very abundant.[241] These orchards had the red scale-insect, *Aonidiella aurantii*—an Asiatic species—living on the trees. By using a suitable insecticide the ants could be killed on some trees without harming the scales. On the trees with ants, scale insects were on the average five times as common; at the peak of the year 150 times. This was because the Argentine ants killed off many though not all of the natural enemies and parasites of the scales. The same kind of thing is very potent in the situation affecting control of the swollenshoot virus disease, one of the major threats to cacao-growing on the Gold Coast. Here two kinds of native insects are involved: *Pseudococcus njalensis* is the commonest scale insect on the trees and spreads the virus by sucking the plant, while ants of the genus *Crematogaster* protect the scale insects.[251] The latter gains by protection and becomes abundant, but also because its honeydew becomes a culture medium for bacteria and moulds that choke the insects if the secretion is not removed by the ants. The ants themselves live in hollows and galleries made by wood-boring insects working in the dead branches of the cacao trees. Another example of the intricacy of relationships that may affect the spread and maintenance of a disease, is seen in the American chestnut blight (described in Chapter 1). It was noticed that one of the commonest places where the fungus entered a tree was the tunnel of a bark or wood-boring insect. 'In many parts of the country where the disease is prevalent there is very direct evidence that bark borers, and particularly the two-lined chestnut borer (*Agrilus bilineatus*), are directly associated in this way with 90 per cent. or more of all cases of this disease.'[259]

So the complexity of natural balance in populations is evident enough to anyone who cares to recognize it. Even when a plant-animal food-chain is transported from its natural region, there may be dislocation because the system, itself simple in structure, has to operate in a new environment. The European larch, *Larix decidua*, occurs naturally in Switzerland, and on it lives a native longicorn beetle, *Tetropium gabrieli* (Fig. 50). Within this natural region there have been no records of the beetles causing damage to the trees. But outside this area, especially in Germany and sometimes in England, damage often develops in planted larches, where the beetle has spread (Fig. 51). Something therefore affects the relationship between tree and insect growing in these new habitats. Gorius found that the trees showed signs of physiological change before the beetles entered them, and that weakening of the trees might therefore be caused by some disbalance with the climate or soil. The weakened trees were for some reason more vulnerable to insect attack.[250]

About the same time, eighty years ago, that the idea of counterpests was being seriously explored, various poisons began to come into vogue for the control of fungus diseases and insect pests of crops. As far as I know, no one has ever produced an effective counterpest for microscopic fungal parasites of plants, though it has recently been discovered that the rust fungus that grows on wheat and grasses has a bacterial parasite and that this in turn may harbour a bacteriophage—a chain of parasite and hyper-parasite.[254] The best antidote for fungal or other parasitic diseases of plants is to find or breed a resistant strain, or let this happen by a rough kind of natural selection. A fungus disease of asparagus in Europe, where it does not develop epidemics, got carried to the United States where it 'swept over the entire country and virtually destroyed the entire asparagus industry. Gradually, however, the rust has become less important until now asparagus growing has become rehabilitated and the disease is of minor importance.'[264] This happy result seems to have taken place by genetic changes in the populations of fungus or asparagus or in both, not by chemical sprays. Though the genetics of such selection are usually complicated, 'commercial breeding usually does not wait for the results of the analysis of the relationship between host and fungus'.[262]

The same thought might be expressed about the incredibly massive use

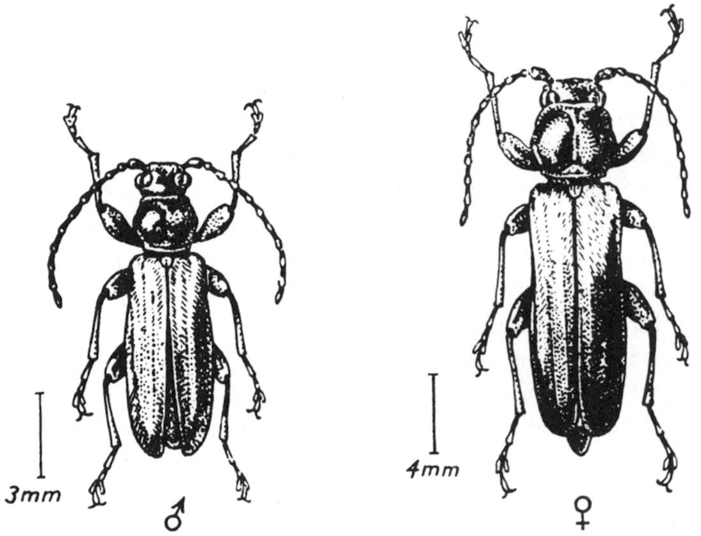

FIG. 50. A longicorn beetle, *Tetropium gabrieli*, whose larvae damage larch outside the natural range of the tree and the beetle. (From U. Gorius, 1955.)

of insecticides now carried on in every part of the crop-growing world. From a state in which only a few well-tried chemicals were in use for this work, we have a yearly increasing number of new ones, and in the last fifteen years especially the new dusts and sprays of the synthetic chlorinated hydrocarbons (which include DDT) and organophosphates (which include parathion). In the same period there has also been a huge expansion in the use of sprays that poison various plants (which include the plant hormones such as 2-4D). With this equipment to destroy parasitic fungi and herbivorous insects and competing weeds, the applied biologist seeks to bypass all the irritating and complex interactions of natural populations, in fact simply sweep away natural food-chains altogether, leaving only the crop plants to give an ordered and useful appearance to the landscape. He is also able to kill the vectors of diseases, such as plant-lice and scale insects that carry and spread the viruses of plants, and the blood-sucking insects that carry diseases of man or domestic animals. In exceptional instances,

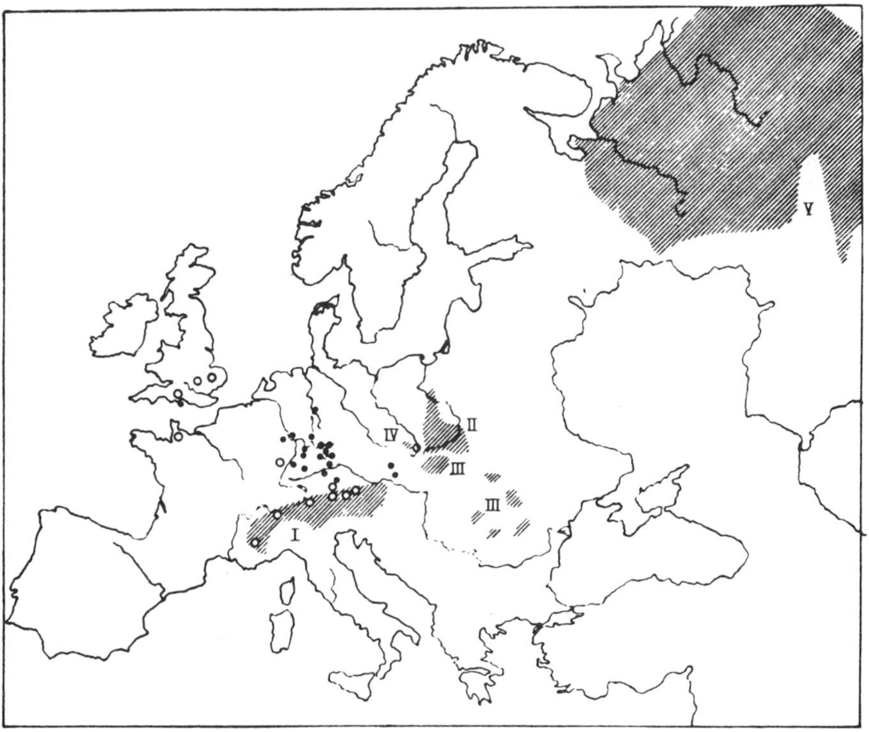

FIG. 51. Distribution of a longicorn beetle, *Tetropium gabrieli*, and its host tree, the larch. The hatched areas I–IV are the natural distribution of the European larch, *Larix decidua*; V is the Siberian larch. The white circles show records of the beetle without damage to the tree, and the black dots instances of outbreaks, the latter being all outside the natural range of the beetle and the tree. (From U. Gorius, 1955. One black dot for England added from Duffy, 1953.)

dusting and spraying do completely obliterate an insect population of some magnitude, as in the Brazil malaria pandemic described in Chapter 1.

There are, I think fortunately for the future of the world's flora and fauna and for man's intelligent appreciation of the world his descendants will live in, two phenomena that work against the complete success of this chemical warfare. The first is the development of 'resistant strains', and the second is compensatory reactions in ecological communities. The populations of certain species of insects that have been systematically poisoned for some years have become less susceptible to particular chemicals, and sometimes to whole related groups of chemicals.[234, 235] Among these species are such star performers as the codlin moth, *Carpocapsa pomonella*, of apple orchards; the black-scale insect, *Saissetia oleae*, and the red-scale insect, *Aonidiella aurantii*, of citrus trees; the cotton aphid, *Aphis gossypii*; and the San Jose scale insect of fruit trees, *Aspidiotus perniciosus*. And as well as in these crop pests resistance has appeared in the house fly, *Musca domestica*, in many countries; in the African blue cattle tick, *Boophilus decoloratus*; in the elm bark-beetle, *Scolytus multistriatus*, in America; and in some malaria mosquitoes. The red scale, which is the chief insect pest of citrus groves in southern California, has become so resistant to fumigation with hydrocyanic acid that trees often get reinfested within a year. In 1941 it was proved that the resistance is inherited in a single sex-linked gene. In later years control has been kept by means of an oil spray instead.

In recent times the chief weapon against malaria mosquito larvae has been poisons distributed often from the air. But *Anopheles gambiae* in Northern Nigeria has produced a strain resistant to dieldrin that shows a simple Mendelian inheritance.[240] Another kind of malaria mosquito, *Anopheles sacharovi*, has developed resistance to DDT in Greece and Lebanon.[253a] Since 1945 salt marshes on eastern Florida have been sprayed with DDT. 'Before this period of treatment salt-marsh mosquitoes occurred in such numbers as to prevent full development of the area. After the treatment, freedom from mosquitoes was considered one of the shining examples of modern insect control measures. Recently the two most prolific insects involved, *Aedes sollicitans* (Wlkr.) and *A. taeniorhynchus* Wied., have been reported as developing resistance to DDT.'[235] These two species do not of course carry malaria. The blue tick is a noteworthy case, because its populations in one part of the South African coast first developed high resistance to arsenical dips; the situation was recovered by using gamma benzene hexachloride (Gammexane) with great success; but resistance towards the second poison has now begun to appear.

It would appear that every insect population is genetically mixed in respect of various characteristics natural to the species, and that in some species these characteristics, such as cuticle structure or enzyme chemistry happen to influence the ability to resist poisons. The violent selection caused by heavy mortality through poisoning leaves the more resistant strains; and there may also be new mutation happening, as is known in bacteria that become resistant to antibiotic drugs. The same process is doubtless at work in populations of parasitic fungi treated by chemical means. Resistance to poisons has only appeared in a few species of arthropods so far, and in them only towards one or a few insecticides, though these are often very important ones. Yet in 1951 Babers and Pratt wrote: 'At present, however, it seems that almost any positive statement concerning resistance will probably have to be rescinded or modified.' [235] We are hearing the early rumblings of what may become an avalanche in strength.

The second drawback of insecticides and fungicides is that they never act only on the single species that is being attacked. They affect the survival of other species like competitors, parasites, and enemies, indeed in some way or other partly alter and may dislocate the population balance of the whole community. Something more is said about this in the next chapter. There are still further chains of effect. 'Heavy annual use of DDT, technical BHC, and probably other persistent chlorinated hydrocarbons appears to have definite danger of reducing within a comparatively few years the productivity of soils to which they are applied.' [246] Experiments in Maryland have proved that DDT, benzene hexachloride, chlordan, and toxaphane (all chlorinated hydrocarbons) depress the growth of seedling crop plants like beans, wheat, and barley. Certain strains of crops also had reduced yields of seed.[239] Such effects need to be set against any immediate saving of insect damage that the poisons may give. However, this is a very new branch of research and obviously a very complex one: little has yet been done to find out all the ways in which these processes may act.[265] The matter is mentioned here just to illustrate how we cannot expect to throw a barrage of selective poisons on to even a fairly simplified ecological system without getting a number of unforeseen effects—effects that should be studied and foreseen before the barrage is ever laid down at all. Experiments have been done to see how soil mites and insects are affected by doses of DDT and other poisons employed in the control of crop pests.[263] The subtlety of the population balance among the small arthropods of cultivated soil was easily realized. BHC (the gamma-isomer of benzene hexachloride) knocked down the numbers of springtails (Collembola) and mites; however, when such plots of ground were simultaneously treated also with DDT the mites decreased but the springtails increased soon afterwards. 'Laboratory tests showed that while Collembola were completely unaffected by DDT, even at the highest concentrations, this substance was definitely toxic to all the Mesostigmata examined.' These Mesostigmata are predacious mites that prey actively on springtails, and their removal had probably taken off the pressure from the latter's populations with a resulting upsurge in numbers. When we remember the really enormous numbers of these insects in any ordinary soil and that they feed largely on the microfungi, it cannot be doubted that the residues of poisons may change the metabolic activity of the soil community and so affect the productivity of crops. Bacterial changes are also involved.

The brief consideration of this astonishing rain of death upon so much of the world's surface is brought in here to prepare the mind for the views developed in the last two chapters of this book, wherein it is suggested that there may be other and more permanent methods of safeguarding the world's organic wealth. No realist would for a moment suppose that either counterpests or chemical warfare can be abandoned, but both can be much modified and adapted to the equal realities of the ecological scene, and the very delicately organized interlocking system of populations that lies within it. Mass destruction and the casual releasing of predators and parasites may some day be looked back upon as we do upon the mistakes of the industrial age, the excesses of colonial exploitation or the indiscriminate felling of climax forests. Aldo Leopold wrote: 'One basic weakness in a conservation system based wholly on economic motives is that most members of the land community have no economic value. Wildflowers and song-birds are examples. Of the 22,000 higher plants and animals native to Wisconsin, it is doubtful whether more than 5 per cent. can be sold, fed, eaten, or otherwise put to economic use. Yet these creatures are members of the biotic community, and if (as I believe) its stability depends on its integrity, they are entitled to continuance.'[255] Before following the practical implications of doing this, we need to examine the reasons and motives for conservation.

CHAPTER EIGHT

The Reasons for Conservation

I once visited a very good school where the headmaster concentrated on getting his pupils interested in running a large vegetable garden. It was a fine garden and the children were obviously enjoying their craft. I asked the master if he had time to tell them anything about animals and he answered: 'Oh yes, I teach them that animals are pests!' This is the understandable point of view of a practical man looking at a limited project; but quite different from that of Robert Browning when he wrote:

I am earth's native:
No rearranging it!

And yet a great literary critic said that Browning's genius had its sound, stubborn roots in real life. It is something to have a point of view towards nature at all. There are over 25,000 different kinds of native land and fresh-water animals in Britain, and probably over a million species of animals in the whole world. The kind of co-existence with them that we can look forward to in the long run depends very much on our attitude to wild life and to nature in general. I think of the human race as being on a very long train journey in company with all these other passengers, and there seem to me to be three absolute questions that sit rather patiently waiting to be answered. The first, which is not usually put first, is really religious. There are some millions of people in the world who think that animals have a right to exist and be left alone, or at any rate that they should not be persecuted or made extinct as species. Some people will believe this even when it is quite dangerous to themselves. Efforts to control plague rats in some Indian warehouses have sometimes been frustrated because the men in charge put out water for the rats to drink. Ideas of this sort will seem folly to the practical Western man, or sentimental. Yet who can really stand up and call them just sentimental when a great scholar and prophet like Dr Schweitzer says: 'The great fault of all ethics hitherto has been that they believed themselves to have to deal only with the relation of man to man'? [287]

The second question can be called aesthetic and intellectual. You can say that nature—wild life of all kinds and its surroundings—is interesting, and usually exciting and beautiful as well. It is a source of experience for poets and artists, of materials and pleasure for the naturalist and scientist. And of recreation. In all this the interest of human beings is decidedly put first.

The third question is the practical one: land, crops, forests, water, sea fisheries, disease, and the like. This third question seems to hang over the whole world so threateningly as rather to take the light out of the other two. The reason behind this, the worm in the heart of the rose, is quite simply the human population problem. The human race has been increasing like voles or giant snails, and we have been introducing too many of ourselves into the wrong places. Consider the hair-raising titles of some fairly recent books about this—serious works, not just written by cranks: *Road to Survival*, *The Rape of the Earth*, *Our Plundered Planet*, *The Geography of Hunger*, *Resources and the American Dream*, *The Limits of the Earth*. Also *The Estate of Man*, in which Michael Roberts suggested that we are reaching the limit of the supplies of inherited talents needed to cope with all these problems. It is just one of the stark facts of this century that man is not only getting more numerous, but wanting more. He is pressing harder than ever in the history of the world into what used to be unexploited, or lightly exploited habitats. And every time he makes a move of this kind, there are new ecological disturbances, including the ones that come from new invasions of plants and animals and their parasites.

So there are the three points of view: you may think the astonishingly diverse life of the globe was not evolved just to be used or abused, and perhaps largely swept away. You may take the view that it is all so interesting and beautiful that it should be preserved, especially preserved for posterity to enjoy. This is not an uncommon attitude in the richer countries, but finds much less favour in those where making a living at all

38. An old wallow of the American bison on the high plains of Kansas, 1899. The last buffalo in Kansas were killed in 1879. (Photo by the U.S. Geological Survey, from M. S. Garretson, 1938.)

39. A huge stack of skulls and other bones of the American bison (perhaps 25,000 in number) at a Saskatchewan railway siding. (Photo by H. Lumsden, 1890, from C. Gordon Hewitt, 1921.)

40 & 41. Complete erosion of topsoil in a farm area in southern Ontario, partly through unrestricted grazing after the original forest was cut. The skeleton stumps of the white pines (*Pinus strobus*) stand bare on their roots. The lower photograph shows on the right new planting and the recolonization of the bare sand by vegetation to reconstitute the land. (Photos by C. S. Elton, 1938.)

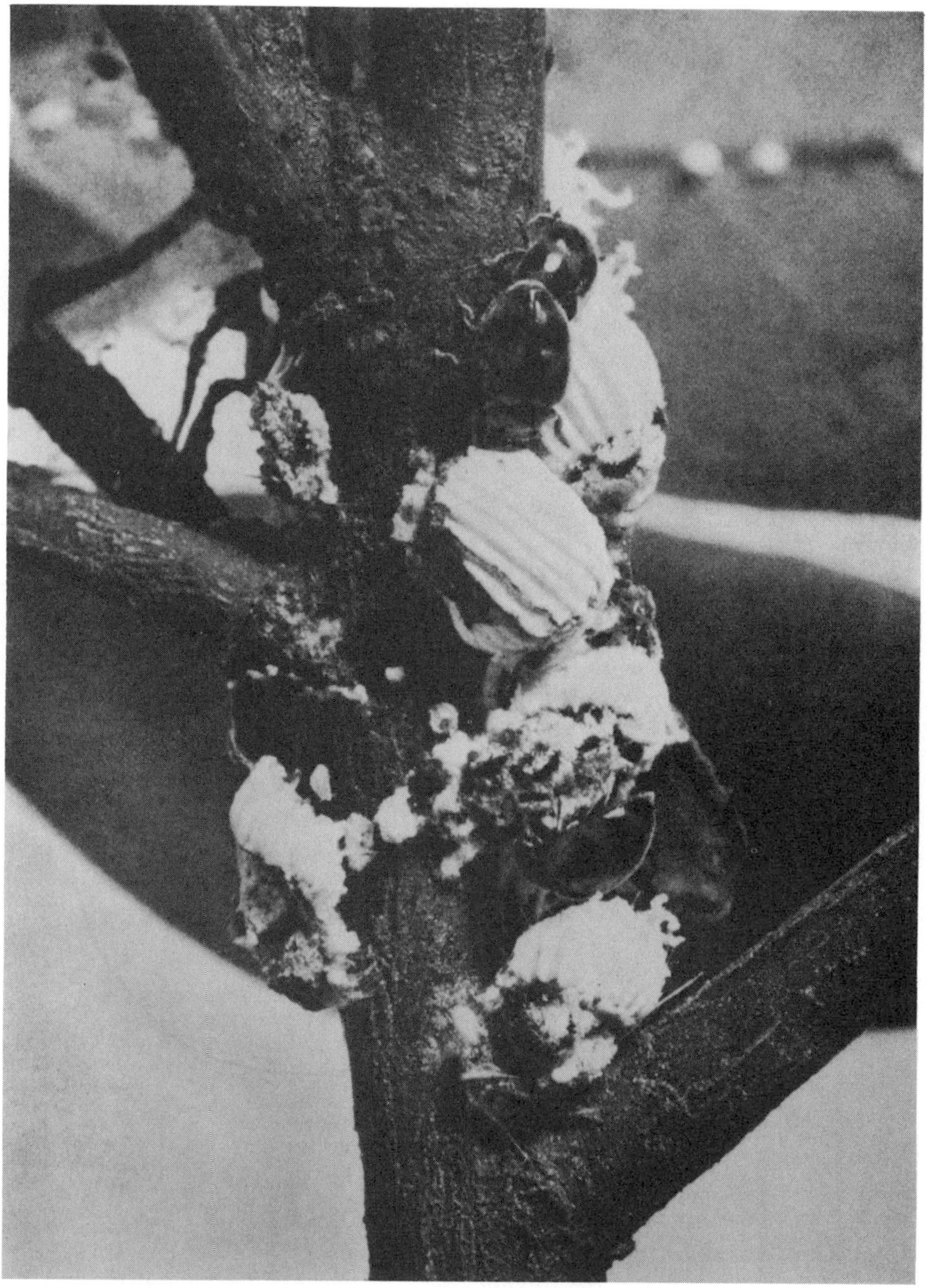

42. The fluted or cottony cushion scale insect, *Icerya purchasi*, attacked by the ladybird, *Novius* (*Vedalia*) *cardinalis*, on a Californian citrus tree. Both the pest and its counter-pest came originally from Australia. (Photo by courtesy of the Department of Biological Control, Citrus Experiment Station, University of California.)

43. Reproduction and mortality: a life-group symbolic of the play of forces in natural control employed in counterpest work. This tachinid fly, *Centeter cinerea*, has darted out and settled for a second or two on the pairing Japanese beetles, *Popillia japonica*, and laid two parasitic eggs on the thorax of the female. The fly was introduced from Japan into the United States in an attempt to control the beetle. (From a painting by a Japanese artist, in C. P. Clausen, J. L. King and C. Teranishi, 1927.)

44. Like the oyster, the big African land snail, *Achatina fulica*, has snail enemies. This one, *Gonaxias*, attacks the body of the snail directly. It has been introduced into the Mariana Islands and from there into Hawaii, in order to try and control the giant snails. (From R. Tucker Abbott, 1951.)

45. A quiet lane in Oxfordshire at the flowering time of cow parsley (*Anthriscus sylvestris*) in May. The hedges are of elm, hazel and maple, with some climbing *Clematis*, and two holly trees left by the hedgers. In summer such a lane is full of flowers and insects. (Photo C. S. Elton, 1957.)

46. A Hampshire roadside at the end of June when scything of the vegetation has just begun. In spite of winds over this chalk hill (shown in the shape of the yew tree) there is a tall mixed meadow along the roadside, with hogweed (*Heracleum sphondylium*) on whose flowers many insects congregate for pollen and nectar. (Photo C. S. Elton, 1954.)

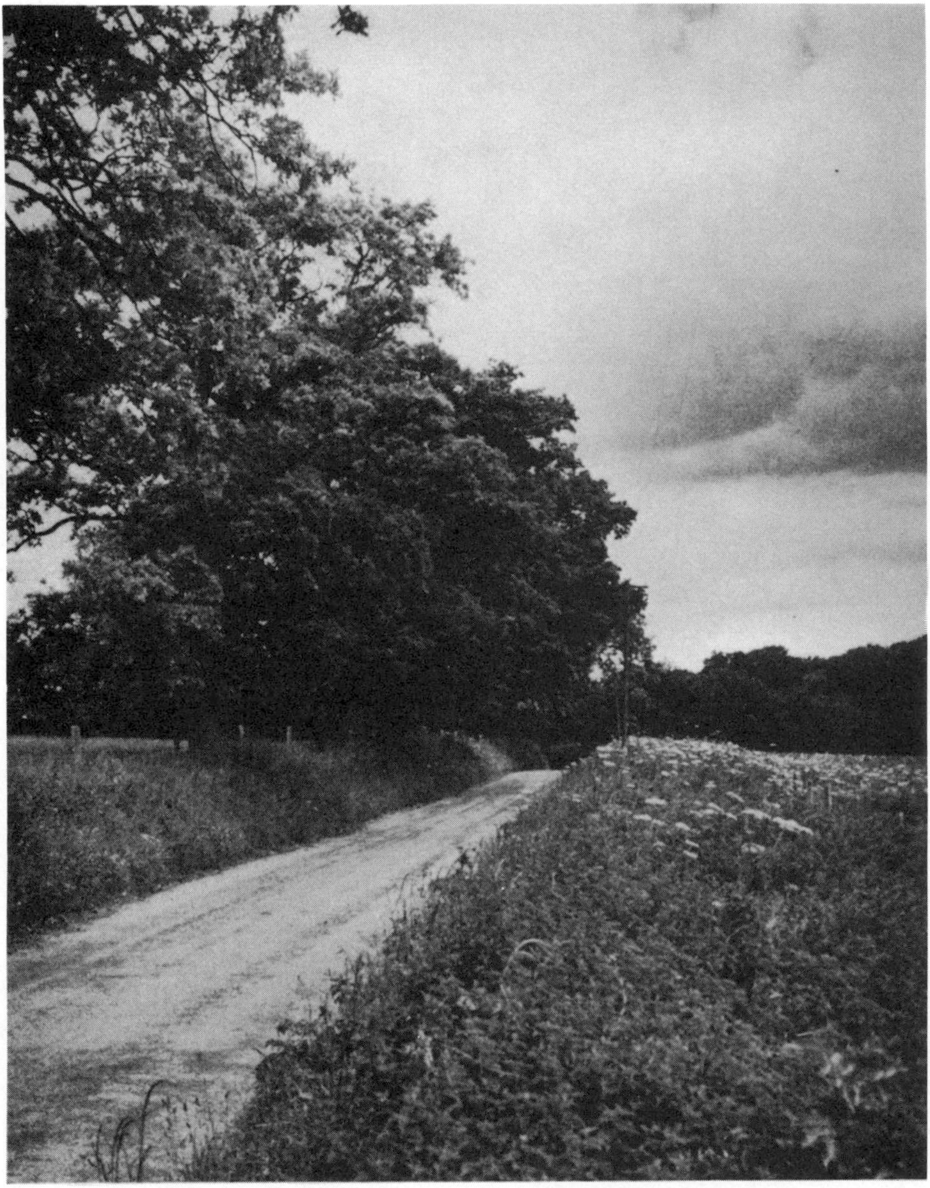

47. Hedgerow trees supply a fifth of British home-grown timber. These oak trees appear to be growing on a grassy bank; but on it are also the remains of a good hedge composed of chalk scrub species, including dogwood, hazel, spindle, rose, blackthorn, and wayfaring tree (*Viburnum*). This Hampshire road was photographed in early June. On the right is a bed of stinging nettle (*Urtica dioica*), a plant that carries an exceptionally rich insect fauna of its own. Behind is a flowering hogweed (*Heracleum sphondylium*). In the centre foreground is a little white dead-nettle (*Lamium album*). (Photo C. S. Elton, 1957.)

48. Aerial view of established contour terraces and hedges designed to check soil erosion on a farm in Indiana. The hedges are composed of *Rosa multiflora* (see next Plate). Several farm ponds can also be seen: nearly two million have been constructed in the United States in the last twenty years, and these contribute to the variety of wild-life. (Photo 1955, by courtesy of the Plant Technology Division, Soil Conservation Service, United States Department of Agriculture.)

49. Five-year-old hedges of flowering *Rosa multiflora* on a farm in Maryland. This introduced Asiatic rose has been planted on a very large scale to make hedges in the Eastern United States. It provides an effective stock fence, harbours wild mammals and birds and insects, and its flowers and fruits are beautiful to look at. (Photo 1954, by courtesy of the Plant Technology Division, Soil Conservation Service, United States Department of Agriculture.)

50. A rich, varied, and balanced ecological pattern. Farm fields, hedgerows, and woods (chiefly of oak and ash with hazel) seen from the top of Midsummer Hill on the Malvern Hills, Herefordshire, where there is an ancient sheep-grazed earthwork. This landscape supports abundant wild life, both plant and animal; many of the species are local and interesting and beautiful; the farming and forestry are healthy and well-managed. Shall we leave this for posterity—of all species? (Photo C. S. Elton, 1953.)

comes first. But wherever you live these practical problems have to be dealt with first. People do have to grow things in order to live and make a living, they need land, and good crops. It is no use pretending that conservation for pleasure or instruction, or the assigning of superior rights to animals will ever take precedence over human survival. Nor should it.

But suppose the conflict between these interests is not quite so great as it seems at first sight? Suppose one could make out a good case for conserving the variety of nature on all three grounds—because it is a right relation between man and living things, because it gives opportunities for richer experience, and because it tends to promote ecological stability—ecological resistence to invaders and to explosions in native populations. This would be a fourth point of view—an attempt to harmonize divergent attitudes. Unless one merely thinks man was intended to be an all-conquering and sterilizing power in the world, there must be some general basis for understanding what it is best to do. This means looking for some wise principle of co-existence between man and nature, even if it has to be a modified kind of man and a modified kind of nature. This is what I understand by *Conservation*. Some time during the next Millennium, when the pressure of human population increase has relaxed, an ecologist may be able to announce that they have pulled this off. Just now it is only possible to give a progress report and a hopeful forecast—the sort of thing a nuclear engineer might have given about power stations ten years ago. Only this ecological forecast is concerned with reducing direct power over nature, not increasing it; of letting nature do some of the jobs that engineers and chemists and applied biologists are frantically attempting now. And the forecast is quite hard-headed, not a sentimental dream.

I will now try to set out some of the evidence that the balance of relatively simple communities of plants and animals is more easily upset than that of richer ones; that is, more subject to destructive oscillations in populations, especially of animals, and more vulnerable to invasions. For if this can be shown to be anywhere near the truth, it will have to be admitted that there is something very dangerous about handling cultivated land as we handle it now, and even more dangerous if we continue to go farther down the present road of 'simplification for efficiency'. It must be remembered that a short précis of evidence can really only introduce a point of view, not prove it. But it is by no means a far-fetched idea, and even if it seemed so we should still need to explore it by research, because the whole matter is supremely important to the future of every species that inhabits the world.

First of all, there are the conclusions of mathematical speculation about population dynamics. One general outcome of one branch of the profound manipulation of very simplified mathematical 'models' of populations, is to bring out the delicacy of balance that may be expected to occur within and between them. The greatest theoretical (and it is only theoretical) discovery has been that populations composed of a single prey species and its enemy, or a single host species and its parasite, would fluctuate in numbers conspicuously, even without shocks from outside like the vagaries of climate. There is, in these simple models, very little inherent stability, but the fluctuations would not necessarily cause extinction. Put in ordinary language, this means that an animal community with only two such species in it would never have constant population levels, but would be subject to periodic 'outbreaks' of each species.

This conclusion from algebra, and especially from the calculus, has been supported to some extent by laboratory experiments done with small animals kept in carefully standardized environments: mostly Protozoa, mites, waterfleas, and beetles. The experiments with Protozoa and mites are those especially apposite here, because in them a single species of prey population is kept with a single species of enemy, though the actual experiments have to be repeated and done in various different ways. One thing again stands out from the results: it is very difficult to keep small populations of this simple mixture in balance, for not only do they fluctuate but one or both of the species is liable to become extinct. The logic of these experiments has not been carried much further, indeed there is still only one important set of them, done by the Russian ecologist G. F. Gause in the nineteen-thirties.[279, 280] Perhaps the technical difficulty of setting up the tests and keeping the population mixtures going has discouraged students from repeating this work. Yet I believe it is fundamentally valuable.

Gause brought another ecological principle into the experiments—

the use of *cover*.[279, 281] By giving cover for the prey to hide and dodge about in, he was able to change its chances of survival. But again, the systems he devised were never stable and either the enemy species died out because it could not get at its food, or else the effectiveness of cover only delayed the final course towards extinction. But in nature the second result might be important, in enabling local immigrants from another part of the system to come in and help to maintain the original population of prey. Indeed, Gause did try introducing a few extra animals from outside, from a nature reserve as it were, and by this means managed to keep the mixture of populations going, albeit still with strong oscillations in numbers. This is just what happens when an orchard is chemically sterilized in the spring, and recolonized from surrounding hedges and woods during the summer (Chapter 9).

The third piece of evidence is that the natural habitats on small islands seem to be much more vulnerable to invading species than those of the continents. This is especially so on oceanic islands, which have rather few indigenous species (Chapter 4). Even our biologically well-populated island of Great Britain has now got about twice as many wild deer and rodent species living in it as there were at the time of the Norman Conquest; and we know that Britain was separated from the Continent some seven thousand years ago before it received its full complement of species on their return north and westwards after the last Ice Age. These additional species have been brought in by man, and some of them have staged considerable outbreaks during the course of their invasions, while some of the woodland deer populations are even now probably building up towards future large eruptions.

The fourth point is that invasions and outbreaks most often happen on cultivated or planted land—that is, in habitats and communities very much simplified by man. They have been simplified in three ways chiefly: by encouraging crops usually of foreign plants that have not a full fauna attached to them, by growing these in partial or complete monoculture, and by trying to kill all other species thought to be harmful, as well as incidentally killing or suppressing a great many more whose fate is not attended to. Furthermore, genetic selection of crop species and also (as mentioned in the last chapter) of some of the pests, upsets the biologically adaptive balance in other ways. Invading species, as has been shown by some of the examples given earlier in this book, do sometimes penetrate natural habitats. This happened with the prickly pear in Australia, the European spruce sawfly in Canada, and the striped bass in Californian seas. The first two of these situations are or look like being rectified by the intelligent introduction of counter-pests—making the community a bit more, not a bit less complicated; the striped bass has not caused any recorded dislocation in the community into which it so suddenly thrust its way. It would, in general, be expected that invaders with unusual ecological power, entering into fairly rich natural communities, will be controllable by ecological methods, or else find a place for themselves without very much disturbance of other populations. However, this again is not at all an invariable rule, as shown by the history of the American grey squirrel in English woods and of the American slipper limpet in oyster beds. In these instances we are presumably seeing some of the 'decisive battles of history'. Nevertheless, it is remarkable that in the actual woodland part of the ecosystem of Wytham Woods (Chapter 6) the rich natural communities have got only three or four prominent invaders from abroad, including the American grey squirrel and the European sycamore, *Acer pseudo-platanus*. The former arrived at Oxford in this century, the latter reached this country several hundred years ago. The chief destructive agent of sycamores is the grey squirrel, which kills the trees by stripping bark off the branches; and this is the only serious cause of damage to them. In this mixture of two invading populations, a balance has still to be attained.

The fifth line of evidence comes from the tropics. It was first brought home to me some years ago, when I had spent an hour expounding ideas about insect outbreaks to three forest officers from abroad. Then one of the men remarked politely that this question did not really concern them, because they do not have insect outbreaks in their forests! I found that he came from British Guiana, another from British Honduras, and the third from tropical India. Now a Dutch forest ecologist, Voûte, believes that this is a general rule about ecological stability in tropical rain forests.[289] Rain forest is very rich in species. His notion is that there are always enough enemies and parasites available to turn on any species that starts being unusually numerous, and by a complex system of checks and buffers, keep them down. Of course this is only a theory, and I expect only part of the story. But the ecological stability of tropical rain forest seems to be a fact. Audy, leader of the British research on medical ecology in Malaya, has shown why scrub typhus has become a dangerous disease there. 'In Malaya the thatch-grass *lalang* (*Imperata cylindrica*), the field-rat or *tikus sawah* (*R. argentiventer*), and one of the scrub-typhus vectors (*Trombicula akamushi*) are dominant species in open wasteland. Not one of these is native to Malaya and wherever it occurs in the deforested areas it does so in denser, more vigorous, populations than you would find of corresponding plants and creatures in the forest . . . the actual infection is probably native to the forest, where however it is practically lost in the complex community—it needed simplification and concentration following destruction of the forest to boost this infection up until it became a danger to man.' [275]

The sixth kind of evidence has been emerging from recent research on orchard pest control. Orchards are specially good for testing the effects of ecological variety, because they are half-way between a natural woodland and an arable field crop—less complex than the wood but more complex than the crop, and more permanent. They are much more drastically managed than woodland, and suffer a great number of foreign invasions by animals and fungi and other pests. Most of these pests have by now reached the orchard countries of the world, so that the whole problem is of tremendous interest to anyone who wants orchard fruit. Here I am more looking at it as an example of man interacting with a relatively simple ecological system.

The most thorough research has been done by a group of Canadian entomologists in Nova Scotia, who have tried to find out the causes of successive waves of pests on apple trees.[284, 286] Several of these are particularly harmful, all originally introduced from abroad:—a fungus, *Venturia inaequalis*, causing the apple scab; the codlin moth caterpillar, *Carpocapsa pomonella*; the oystershell scale insect, *Lepidosaphes ulmi*; and the European red mite, *Metatetranychus ulmi*. From about 1930 onwards a puzzling series of outbreaks began to blow up. The extraordinary discovery was made that these were almost certainly caused by side effects of a fungicide spray upon the enemies and parasites of the animals.[282] A change in chemical composition of one of these sprays used against apple scab was followed by enormous multiplication of scale insects on the apple bark and twigs. It was found that the old spray killed the scale insects, one of its enemies and one of its parasites. But the new one left the scale insects unharmed, while still destroying its enemy and parasite, thus proving again the value of the old Chinese proverb that 'there is no economy in going to bed early to save candles if the result be twins'.

Other peculiar results of spraying have also come to light. In recent years DDT has been used as an insecticide in orchards all over the world, partly in the control of the codlin moth. But it turns out that this kills the enemies of the European red mite without being a control of the mite itself.[283] There has therefore arisen a worldwide abundance of red mites in orchards. Collyer has found that there are at least forty-five species of insects and mites that prey upon the European red mite in Essex orchards.[276, 277] Both here and in Canada neglected orchards have very low red mite populations and a good mixture of natural enemies—for they do not all prey exclusively on this one pest, but have a range of natural prey. No one, however, imagines that apple orchards could safely be left to find their natural balance and all spraying be stopped. But it is evidently a very touchy problem how to maintain a balance artificially, and one leading Canadian orchard ecologist has remarked that 'We move from crisis to crisis, merely trading one problem for another'.[285]

The six lines of evidence just given can be summarized as follows. Mathematical concepts about the properties of the food-chain, and simplified laboratory experiments, prepare our minds for instability in very simple population systems. In them we may expect strong oscillations and often extinction. If the habitat is given additional structural properties in the form of cover, there may be some mitigation of this instability, though complete success in experiment is still very rare. Oceanic islands and crop monocultures are simple ecosystems that show high vulnerability to invasions (whether from other lands or from other habitats in the same country) and frequent outbreaks of population subsequently. But tropical rain forest has these features damped down to a remarkable degree. An orchard that has not been treated with insecticide achieves an ecological stability amongst its hundred or more species of animals, though it does not reach the standards of quality and abundance of fruit that are wanted. The explosions of pests in orchards have partly been due to new invasions from without, partly to the numerous accidents and interactions that affect any animal community, but in a notable degree to upsetting of the relationships between pests and their natural enemies and parasites through differential effects of the poisons used. These six lines of evidence all seem to converge in the same direction, though each of them really requires much more extensive analysis and discussion than can be given here. The argument is put forward because, if it is correct, there is a prospect of being able to handle our biological affairs by the better planning of habitat interspersion and the building up of fairly complex plant and animal communities, as I shall explain in the next chapter. And we should try to do this for the three reasons already given: in order to create refuges for wild species, in order to make our surroundings interesting and satisfying, and in order to promote stability of populations and a varied community in which all kinds of compensatory pressures will be exercised on populations.

These arguments are not at all intended to promote any idea of complete *laissez faire* in the management of the ecosystems of the world. The breakdown of Wallace's Realms, the outburst of human populations and the advances of technological power have put an end to any idea like that. The world's future has to be managed, but this management would not be just like a game of chess—more like steering a boat. We need to learn how to manipulate more wisely the tremendous potential forces of population growth in plants and animals, how to allow sufficient freedom for some of these forces to work amongst themselves, and how to grow environments—for example, certain kinds of cover—that will maintain a permanent balance in each community. I think that in pest control there is already a turn of the tide in ideas. Two examples may be given. As mentioned in Chapter 7, the worst pest of citrus orchards in southern California is the Asiatic red scale insect, *Aonidiella aurantii*, which has to be attacked regularly by spraying various insecticides. Introduction of counterpests had been tried but not found very effective. The Division of Biological Control of the University of California Citrus Experiment

Station began an interesting experiment in 1948.[278] Of seventeen lemon and orange groves that were not having any spraying treatment, three had had none for one or two years past, and five more no treatment for at least eight years. After making sure that there would not be conflicting results from different strains of trees or from diseases, attention was concentrated on the numbers of predators and parasites in each grove. Protection of scale insects on whole branches of a tree, so that they could have no enemies or parasites, proved that the latter were very important in controlling the pest. Sprayed trees quickly built up an abundance of scale insects again, whereas in the absence of spraying this was much less marked. The chief agent of control was an Asiatic parasite, the golden chalcid, *Aphytis chrysomphali*, already widely spread in California at this date. Although the percentage of hosts carrying the larvae of the parasite was not particularly high, the adult insects were found to feed on the scales themselves and to account for much of the additional mortality. There are also three predatory insects that become abundant when the scales increase, but these cannot achieve control by themselves, at any rate under the particular local conditions. The report on this work states that 'annual insecticidal treatment for the red scale probably precludes the possibility of attaining satisfactory natural balance'.

Once the notion is grasped that complexity of populations is a property of the community, to be studied and used in conservation, there is hardly any limit to the ways in which it could be introduced. The same group of Californian research workers has tried out another idea that at first sight seems most surprising.[288] The black scale insect, *Saissetia oleae*, also an invader from Asia, is another serious pest of citrus orchards. It is particularly abundant in the inland parts of the State, because the more continental type of climate there brings about each season a sharp cycle of abundance in the scale insects and its common chalcid parasite, *Metaphycus helvolus*. On the coastal areas there is a longer season and more overlapping of generations, which spreads out the effects of the parasite. In the first situation there are sharp fluctuations and when the host is scarce the parasite declines so far in numbers that it cannot quickly climb up again. But it was thought that if a species of nightshade, *Solanum douglasii*, that often grows under citrus trees could be preserved there the scale insects would develop earlier because they grow faster on this plant; thus the availability of hosts would be extended further round the season. Previously the nightshade had been looked upon as a detrimental weed for this very reason—that it is an alternative host-plant for the scale insect. These particular experiments 'were terminated by the accidental removal of most of the nightshade plants late in July by zealous, but uninformed orchard workers'. This is not the whole range of the experiments, which involved careful timing and also a certain amount of artificial infesting of the nightshade plants, as also of the citrus trees. The reason they are good experiments is not that they had at that time succeeded in proving their practical use, but that they were exploiting the principles upon which nature actually works, not the principles upon which an engineer or a chemist works.

It is a very long haul from handling a small group of four species like the lemon tree, the nightshade, the black scale, and a chalcid parasite, to the contemplation of the almost inconceivable and profuse richness of a tropical rain forest, or even to the several thousand species living in Wytham Woods, Berkshire. It is a question for future research, but an urgent one, how far one has to carry complexity in order to achieve any sort of equilibrium. Underlying the whole of this issue is also the question of the rate of genetic adjustments in species. The appearance of resistant strains of insects and ticks after a relatively few years of poison treatment proves that this can be quite rapid. So does the spread of deliberate breeding of strains of plant or animal resistant to disease or to some insect pest. Shall we see similar adjustments between counterpests and their hosts or prey? This is not unlikely, and again brings out the necessity of allowing play to the whole power of a community through various channels of biotic pressure. Even if some special relationships become genetically changed, the others might remain for a long time. This idea leads to the question of how to explore all methods of keeping or creating sufficiently rich plant and animal communities in our changing landscape—that is, of conserving ecological variety.

CHAPTER NINE

The Conservation of Variety

'And the vessel that he made of clay was marred in the hands of the potter; so he made it again another vessel, as seemed good to the potter to make it.'

Jeremiah xviii. 4

Except in most of the ocean, the wildernesss is in retreat, or is being changed in character. In another thousand years most of the world's surface and much of its fresh water will have been altered and fashioned by man, or at any rate covered with living communities of plants and animals profoundly different from what they were even a few hundred years ago. Even where the outer form of vegetation has preserved as emblance of primitive character (as so often in forest after it recovers from fire or shifting cultivation), we must grow accustomed to the idea that its plant and animal populations (the latter mostly hidden from casual sight) will have changed their composition and their intricate structure and relationships. And all the time these communities will continue to be invaded by the species arriving from other parts of the world. So far the brunt of these invasions has been borne by the communities much changed and simplified by man. But some invaders are also penetrating the more stable and mature communities of ocean and natural forest. On the exploited lands of the world we see a decrease in richness and variety of species: monocultures with rigid spraying programmes, pastures of pure grass populations, pure stands of trees, the replacement of stratified and mature deciduous woods by quick-growing conifers with their relatively barren structure and poor inhabitants, the cleaning up of waste patches, the hormone spraying of roadsides, and the planting of exotic species many of which may literally be quite sterile of animal life—at first. We might sum up this stream of events in the words of Isaiah:

'Woe unto them that join house to house, that lay field to field, till there be no place, that they may be placed alone in the midst of the earth!'

But man cannot live on gloom alone. There are a great many ways in which the countries we live in could be made more safe for wild life, more interesting, and also more secure for the farmer, forester, and fisherman. If the wilderness is in retreat, we ought to learn how to introduce some of its stability and richness into the landscapes from which we grow our natural resources. Conservation is a protean word, for it can mean on the one hand the preservation of wild species against the advance of human exploitation; alternatively, the methods of attaining the highest productivity from exploited lands. We need to be clear what kind of conservation is meant when it is talked about. If the lines of argument developed in this book are sound, I believe that conservation should mean the keeping or putting in the landscape of the greatest possible ecological variety—in the world, in every continent or island, and so far as practicable in every district. And provided the native species have their place, I see no reason why the reconstitution of communities to make them rich and interesting and stable should not include a careful selection of exotic forms, especially as many of these are in any case going to arrive in due course and occupy some niche.

It will be easier to understand what may seem a rather vague idea, if an example is given. Much of our own highly managed landscape is still interlaced with a wonderful network of hedgerows and roadside verges. These long winding strips of habitat by the road and lane and field margins are the last really big remaining nature reserve we have in Britain, except for the wild moors and lakes of our northern mountains and the seas around us. We need plenty of smaller nature reserves for special purposes, to help some animal or plant or kind of habitat to survive. But our roadside hedges and verges are unique, because they run for something like 190,000 miles amongst our cultivated land and part of our urban land too. A far greater length also borders the mosaic of fields. Hedges are taken for granted by most people. In so many parts of the country they are implicit in the natural scene (Pls. 45, 46). One of the best things Richard Jeffries ever wrote was a simple description of a hedge leading up to chalk downs.[295] Karel Čapek once said: 'I have wandered along roads lined with quickset hedges, sheer quickset hedges which make England the real England, for they enclose, but do not oppress.'[290] One can still find plenty of good hedgerows, most of them of hawthorn and associated scrub and some trees, carefully managed in a rotation of cutting; but many of them are beginning to vanish under what I can only call the engineer's dream of agriculture. A great motor manufacturer recently said in one of our farming magazines: 'On the constructive side let me first emphasize the need for us to pull up some of our hedges.' You can tell that he was not thinking about the conservation of ecological variety.

Several years ago the roadside meadows of our country roads were threatened with a mass attack by chemical herbicides that kill off many of the flowers and leave grass. This campaign might have got under way without the ordinary person being consulted, without asking whether it mattered that we should lose the blue meadow geranium or other beautiful flowers along the roadside, whether there would be peculiar side effects like those in orchards when you spray fungus or insect (Chapter 8). One spraying firm included the wild rose as one of the weeds to be destroyed. Fortunately the Nature Conservancy, whose job it is to take a long view where the country's ecology is concerned, was able to postpone this threat by appealing to the common sense of Government departments, agricultural entomologists, commercial spraying firms, and county road engineers. The campaign was partly suspended while some botanical research was done on the problem, and this helped to produce second thoughts in the operators. At the present time, a rather intelligent compromise seems to have been reached, whereby a strip is sprayed or more usually mown (now often by mechanical cutting), leaving behind a natural garden of herbage in the front of the field boundary.

The hedge and road meadow verge are really extraordinarily variable in structure and communities. No stretch of roadside is quite like any other. But nearly all are ecologically rich, usually stable, except where road repairs and so on make a temporary disturbance—which only adds to the variety of ecological succession that can be seen. Would it not be worth considering that we have here one of our most priceless properties—much of it owned by the nation or by its local authorities, though also by the man on the other side of the hedge. I cannot think of any ecological system in Britain that so clearly has all the virtues inherent in the conservation of variety. There is a refuge for wild life: small nesting birds, and abundant insect populations, not only on the leaves and twigs, but visiting the flowers in early summer. A great many of the species are ones that also live at the edge of a wood or a woodland glade—a habitat now increasingly sterilized by modern forestry managements. Then there is extreme pleasure for the traveller—the flowering hawthorn hedge and its associated shrubs like dogwood and rose and elder, the roadside flowers and the insects that frequent them—like the brimstone butterfly that breeds on the buckthorn shrub. I know a Danish family that came to England especially for their summer holiday in order to see and photograph along our roadsides wild flowers which have disappeared from parts of Denmark through intense cultivation and the use of herbicides. For the naturalist and ecologist there are, besides these pleasures, the fascination of habitats in which may be found perhaps as many as half our British plant and animal species. Next to woodland, it is likely that a fully developed herb-grass meadow is one of the richest communities we have.

There is also a surprisingly full list of economic reasons for the keeping of hedgerows, as also of roadsides themselves, apart from the use of the former as fences and the latter as footways and safety zones for vehicles. The older idea that roadsides are a reservoir of weeds is rapidly coming to have little meaning, as crops become more and more elaborately sprayed. Little ecological research has been done on hedges in this country, and the only general map of the distribution of different kinds of field boundary in England was compiled by a German geographer.[294] Although we commonly think of a hedge as giving shelter, it is not generally known that in Schleswig Holstein their existence has been estimated to increase grain yields by 20 per cent.; and to reduce evaporation caused by wind from the surface of fields by an amount equivalent to having one third more rainfall.[296] Hedges are also a source of timber there, considered equally productive with afforested woodland. And in Britain we get a fifth of our homegrown timber supply from hedgerow trees, mostly oak, elm, and ash (Pl. 47).[292] The trees also give shade for domestic animals and people. How far these strips of habitat form a reservoir for the enemies and parasites of insects and mite pests of crops, is a subject for future research.

But one recent study shows that this is more than a theoretical notion.[291] A comparison of the spiders found in sprayed and unsprayed fruit orchards in Essex and Kent proved that spraying knocks out the spider populations that normally live in an unsprayed orchard, but that after spraying is over there is an infiltration of spiders from the woods and hedges round about. Twenty species were found in sprayed and forty-one in unsprayed orchards. The practical point of this work is that some of these spiders prey on the mites that attack the fruit trees, and that the number of individual spiders in unsprayed orchards was something like twenty-five times greater than in sprayed ones, and but for colonization of the latter from outside there would have been almost none. How far this actually matters for pest control on fruit trees is not yet proved.

Hedges are a fairly recent addition to the British landscape, mostly through enclosure of fields and the development of roads in the last two or three hundred years. Yet the fact that they have rich and stable communities, containing much of the flora and fauna both of meadowland and of woodland edge, and some of woodland itself, proves that it is possible to create and interweave such well-tenanted strips of habitat effectively amongst the more severely exploited fields and woods. They form, as it were, a connective tissue binding together the separate organs of the landscape. To them we should add also railway embankments; though these are not accessible, they provide valuable refuges and a great deal of visual pleasure to the traveller. They are different from roadsides in being partly the result of a great deal of burning succession. Another main network of habitat that should be mentioned is the inland water channels of the country. There are about 1,500 miles of canals, not counting the equal amount of rivers used for that purpose; and a far greater length of river and stream and ditch, all of which contribute to the interest and depth of country ecology.

The examples just given lead on to the idea of actually planning a better and more varied landscape, bringing in all the considerations that affect conservation. In no country has this been attempted with such remarkable drive and imagination as in the United States, where the spur for action has been soil erosion, combined with a fervent interest in preserving habitats for wild game. Edward Graham has given the best general picture of this work and its background of ecological ideas, in his *Natural Principles of Land Use.*[293] A recent summary by the same author[293a] records the extraordinary progress in land diversification made in parts of that country in the last decade, at the same time pointing out how much further ecological knowledge is needed in order to carry through plans of this kind safely. He says: 'There should be additional evaluations of ecological inter-relationships on areas devoted exclusively to nature and on areas under other types of use, as farming, ranching and forestry.' Much of the work of the United States Soil Conservation Service is concerned with putting back what had been lost, or creating entirely new kinds of habitat interspersion (Pls. 48, 49), whereas in Britain we might still have the chance of keeping our own remarkable landscape before it loses its ecological variety. This landscape pattern was built up by individuals, mostly country people working by instinct and making a place to live in, not just a place to raise cash crops in. It can still be seen at its best, for example, in Herefordshire (Pl. 50). From now on, it is vital that everyone who feels inclined to change or cut away or drain or spray or plant any strip or corner of the land should ask themselves three questions: what animals and plants live in it, what beauty and interest may be lost, and what extra risk changing it will add to the accumulating instability of communities. That is: refuge, beauty and interest, and security. This outlook may enable us to put into the altered landscape some of the ecological features of wilderness. Would it not be good to be able to say, like John Muir, the Scotsman who became the great American prophet of wilderness conservation: 'To the sane and free it will hardly seem necessary to cross the continent in search of wild beauty, however easy the way, for they find it in abundance wherever they chance to be.'[297] Will we be able to talk like this in fifty years' time, as he could do fifty years ago?

References

CHAPTER I

1. ARTIMO, A. 1949. [Finland a profitable muskrat land. Preliminary report.] *Suom. Riista*, **4**: 7–61.
2. CABRERA, A., and YEPES, J. 1940. *Historia natural ediar. Mammiferos Sud-Americanos (vida, costumbres y descripcion)*. Buenos Aires.
3. COOKE, M. T. 1928. The spread of the European starling in North America (to 1928). *Circ. U.S. Dep. Agric.* **40**: 1–9.
4. DAVIS, D. H. S. 1953. Plague in Africa from 1935 to 1949. A survey of wild rodents in African territories. *Bull. World Hlth Org.* **9**: 665–700.
5. DE VOS, A., MANVILLE, R. H., and VAN GELDER, R. G. 1956. Introduced mammals and their influence on native biota. *Zoologica, N.Y.* **41**: 163–94.
6. DORST, J., and GIBAN, J. 1954. Les mammifères acclimatés en France depuis un siècle. *Terre et la Vie*, **101**: 217–29.
7. EAST, B. 1949. Is the lake trout doomed? *Nat. Hist., N.Y.* **58**: 424–8.
8. FORESTRY COMMISSION. 1950. Chestnut blight caused by the fungus *Endothia parasitica. Bookl. For. Comm.* **3**: 1–6.
9. FREEMAN, R. B. 1946. *Pitrufquenia coypus* Marelli (Mallophaga, Gyropidae), an ectoparasite on *Myocastor coypus* Mol. *Ent. Mon. Mag.* **82**: 226–7.
10. GRAVATT, G. F., and MARSHALL, R. P. 1926. Chestnut blight in the Southern Appalachians. *Dep. Circ. U.S. Dep. Agric.* **370**: 1–11.
11. HARRIS, V. T. 1956. The nutria as a wild fur mammal in Louisiana. *Trans. 21st N. Amer. Wildl. Conf.*: 474–86.
12. HILE, R., ESCHMEYER, P. H., and LUNGER, G. F. 1951. Decline of the lake trout fishery in Lake Michigan. *Fish. Bull., U.S. Fish & Wildlife Service*, **52** (No. 60): 77–95.
13. HOESTLANDT, H. 1945. Le crabe chinois (*Eriocheir sinensis* Mil. Ed.) en Europe et principalement en France. *Ann. Epiphyt.* N.S. **11**: 223–33.
14. HUBBARD, C. E. 1954. *Grasses*. Harmondsworth, Middlesex.
15. JOHNSON, C. E. 1925. The muskrat in New York: its natural history and economics. *Roosevelt Wild Life Bull.* **24**: 193–320.
16. JOHNSON, SAMUEL. 1775. *A journey to the western islands of Scotland.* London.
17. KALMBACH, E. R. 1954. Pigeon, sparrow and starling control. *Pest Control*, **22** (5): 9–10, 12, 31–2, 54.
18. KUZNETZOV, B. A. 1944. [VIII. Order Rodents, Ordo Rodentia.] In BOBRINSKII, N. A., KUZNETZOV, B. A., and KUZYAKIN, A. P. [*Key to the mammals of the U.S.S.R.*] Moscow. (In Russian.)

19. LANGLOIS, T. H. 1954. *The western end of Lake Erie and its ecology*. Ann Arbor.
20. LAURIE, E. M. O. 1946. The coypu (*Myocastor coypus*) in Great Britain. *J. Anim. Ecol.* **15**: 22–34.
21. LENNON, R. E. 1954. Feeding mechanism of the sea lamprey and its effect on host fishes. *Fish. Bull., U.S. Fish & Wildlife Service*, **56** (No. 98): 247–93.
22. LINK, V. B. 1955. A history of plague in the United States of America. *Publ. Hlth Monogr.*, Wash. **26**: 1–120.
23. METCALFE, H., and COLLINS, J. F. 1911. The control of the chestnut bark disease. *Farmers' Bull. U.S. Dep. Agric.* **467**: 1–24.
24. MEYER, K. F. 1942. The known and the unknown in plague. *Amer. J. Trop. Med.* **22**: 9–36.
25. MYERS, J. G. 1934. The arthropod fauna of a rice-ship, trading from Burma to the West Indies. *J. Anim. Ecol.* **3**: 146–9.
26. PETERS, N., and PANNING, A. 1933. Die chinesische Wollandkrabbe (*Eriocheir sinensis* H. Milne-Edwards) in Deutschland. *Zool. Anz.* **104** (Suppl.): 1–180. (Abstract by C. Elton (1936) in *J. Anim. Ecol.* **5**: 188–92.)
27. SHEAR, C. L., STEVENS, N. E., and TILLER, R. J. 1917. *Endothia parasitica* and related species. *Bull. U.S. Dep. Agric.* **380**: 1–82.
28. SOPER, F. L., and WILSON, D. B. 1943. Anopheles gambiae *in Brazil 1930 to 1940*. New York.
29. TANSLEY, A. G. 1939. *The British Islands and their vegetation*. Cambridge.
30. ULBRICH, J. 1930. *Die Bisamratte: Lebensweise, Gang ihrer Ausbreitung in Europa, wirschaftliche Bedeutung und Bekämpfung*. Dresden.
31. ULM, A. 1948. The Chinese chestnut makes good. *Amer. Forests*, **54**: 491, 518, 520 and 522.
32. VERESHCHAGIN, N. K. 1941. [Establishment of the nutria (*Myocastor coypus* Mol.) in west Georgia.] *Trav. Inst. Zool. Acad. Sci. R.S.S.G.* **4**: 3–42. (In Russian.)

CHAPTER II

33. ARKELL, W. J. 1956. *Jurassic geology of the world*. Edinburgh and London.
34. CAMPBELL, D. H. 1944. Relations of the temperate floras of North and South America. *Proc. Calif. Acad. Sci.* **25**: 139–46.
35. DAMMERMAN, K. W. 1948. *The fauna of Krakatau*. Amsterdam.
36. DARWIN, C. 1845. *Journal of researches into the natural history and geology of the countries visited during the voyage round the world of H.M.S. 'Beagle'*. London.
37. DE CHARDIN, P. T. 1940. The movements of the fauna between Asia and North America since the Lower Cretaceous. *Proc. 6th Pacif. Sci. Congr. 1939*, **3**: 647–8.
38. EKMAN, S. 1953. *Zoogeography of the sea*. London.
39. HICKSON, J. S. 1889. *A naturalist in North Celebes*. London. p. 190.

40. HOPWOOD, A. T. 1953. In MATTINGLEY, P. F. Distribution of animals and plants in Africa. *Nature*, **171**: 639–40.
41. LEAKEY, L. S. B., and CLARK, W. E. LE GROS. 1955. British-Kenya Miocene Expeditions. *Nature*, **175**: 234.
42. LEE, SHUN-CHING. 1935. *Forest botany of China*. Shanghai.
43. MAYR, E. 1944. Wallace's Line in the light of recent zoogeographic studies. *Quart. Rev. Biol.* **19**: 1–14.
44. MAYR, E. 1946. History of the North American bird fauna. *Wilson Bull.* **58**: 3–41.
45. MURPHY, R. C. 1926. Oceanic and climatic phenomena along the west coast of South America during 1925. *Geogr. Rev.* **16**: 26–54.
46. MYERS, G. S. 1953. Paleogeographical significance of fresh-water fish distribution in the Pacific. *Proc. 7th Pacif. Sci. Congr. 1949*, **4**: 38–48.
47. NORMAN, J. R. 1931. *A history of fishes*. London.
48. RAVEN, H. C. 1935. Wallace's Line and the distribution of Indo-Australian mammals. *Bull. Amer. Mus. Nat. Hist.* **68**: 177–293.
49. RENSCH, B. 1936. *Die Geschichte des Sundabogens: eine tiergeographische Untersuchung*. Berlin.
50. SCHAEFFER, B. 1953. The evidence of the fresh-water fishes. *Bull. Amer. Mus. Nat. Hist.* **99**: 227–34.
51. SCOTT, W. B. 1937. *A history of land mammals in the Western Hemisphere*. New York.
52. SIMPSON, G. G. 1953. *Life of the past. An introduction to palaeontology*. New Haven.
53. STOCK, C. 1942. A ground sloth in Alaska. *Science*, **95**: 552–3.
54. VAN DYKE, E. C. 1940. The origin and distribution of the coleopterous insect fauna of North America. *Proc. 6th Pacif. Sci. Congr. 1939*, **4**: 255–68.
55. WALLACE, A. R. 1860. On the zoological geography of the Malay Archipelago. *J. Linn. Soc. (Zool.)* **4**: 172–84.
56. WALLACE, A. R. 1869. *The Malay Archipelago: the land of the orang-utan and the bird of paradise. A narrative of travel with studies of man and nature*. London.
57. WALLACE, A. R. 1876. *The geographical distribution of animals*. London. 2 vols.

CHAPTER III

58. ADAMS, J. A. 1949. The Oriental beetle as a turf pest associated with the Japanese beetle in New York. *J. Econ. Ent.* **42**: 366–71.
59. BAKER, W. A., and VANCE, A. M. 1938. Status of the European corn borer in 1937. *J. Econ. Ent.* **31**: 348–53.
60. BALCH, R. E. 1939. The outbreak of the European spruce sawfly in Canada and some important features of its bionomics. *J. Econ. Ent.* **32**: 412–18.
61. BEZANT, E. T. 1956. Further records of the Australian carpet beetle, *Anthrenocerus australis* (Hope) (Col., Dermestidae) in Britain. *Ent. Mon. Mag.* **92**: 401.

61*a*. BIRD, F. T., and ELGEE, D. E. 1957. A virus disease and introduced parasites as factors controlling the European spruce sawfly, *Diprion hercyniae* (Htg.), in central New Brunswick. *Canad. Ent.* **89**: 371–8.

62. BREWER, E. G. 1941. The fight for the elms. *Amer. Forests*, **47**: 22–5.

63. BROWN, A. W. A. 1941. Foliage insects of spruce in Canada. *Tech. Bull. Dep. Agric. Can.* **31**: 1–29.

64. BUTCHER, D. 1941. Your shade trees and the Japanese beetle. *Amer. Forests*, **47**: 396–7.

65. CHITWOOD, B. G. 1951. The golden nematode of potatoes. *Circ. U.S. Dep. Agric.* **875**: 1–48.

66. CLAUSEN, C. P., KING, J. L., and TERANISHI, C. 1927. The parasites of *Popillia japonica* in Japan and Chosen (Korea) and their introduction into the United States. *Dep. Bull. U.S. Dep. Agric.* **1419**: 1–55.

67. COLLINS, C. W. 1938. Two elm scolytids in relation to areas infected with the Dutch elm disease fungus. *J. Econ. Ent.* **31**: 192–5.

68. COMMONWEALTH INSTITUTE OF ENTOMOLOGY. *Distribution maps of insect pests.* London.

69. COPPELL, H. C., and ARTHUR, A. P. 1953. Notes on introduced parasites of the European pine shoot moth, *Rhyacionia buoliana* (Schiff.) (Lepidoptera: Tortricidae), in Ontario. *84th Rep. Ent. Soc. Ont.*: 55–8.

70. COPPEL, H. C., and LEIUS, K. 1955. History of the larch sawfly, with notes on origin and biology. *Canad. Ent.* **87**: 103–11.

71. CRAIGHEAD, F. C. *et al.* 1949. Insect enemies of Eastern forests. *Misc. Publ. U.S. Dep. Agric.* **657**: 1–679.

72. CRESSMAN, A. W., and PLANK, H. K. 1935. The camphor scale. *Circ. U.S. Dep. Agric.* **365**: 1–19.

73. DE VOS, A., MANVILLE, R. H., and VAN GELDER, G. 1956. Introduced mammals and their influence on native biota. *Zoologica, N.Y.* **41**: 163–94.

74. DOWDEN, P. B. 1939. Present status of the European spruce sawfly, *Diprion polytomum* (Htg.), in the United States. *J. Econ. Ent.* **32**: 619–24.

75. EUROPEAN AND MEDITERRANEAN PLANT PROTECTION ORGANIZATION. 1956. *Leptinotarsa decemlineata* Say, Colorado beetle, Europe—1955. Paris.

76. EUROPEAN AND MEDITERRANEAN PLANT PROTECTION ORGANIZATION. 1957. *Hyphantria cunea* Drury. *Report of the Fourth International Conference on fall webworm.* Paris.

77. EUROPEAN PLANT PROTECTION ORGANIZATION. 1953. *Colorado beetle in Europe in 1952.* Paris.

78. FAIRCHILD, D. 1945. *Garden islands of the Great East: collecting from the Philippines and Netherlands India in the Junk 'Cheng Ho'.* New York.

79. FOREST INSECT AND DISEASE SURVEY, DEPT. OF AGRICULTURE, CANADA. 1956. *Annual Rep. 1955*: 9 and 21.

80. FOREST INSECT SURVEY, DEPT. OF AGRICULTURE, CANADA. 1943. *Annual Rep. 1942*: 3–6.

81. FOREST INSECT SURVEY, DEPT. OF AGRICULTURE, CANADA. 1949. *Annual Rep. 1948*: 11 and 26.

82. FOREST INSECT SURVEY, DEPT. OF AGRICULTURE, CANADA. 1951. *Annual Rep. 1950*: 9.

83. FORESTRY COMMISSION. 1938. Elm disease. *Leafl. For. Comm., Lond.* **19:** 1–8.

84. FORTE, P. N. 1956. The eradication of pests: some observations on the Argentine ant campaign in Western Australia. *J. Dep. Agric. W. Aust.* **5,** Ser. 3, (5): 1–8.

85. GAMBRELL, F. L., MENDALL, S. C., and SMITH, E. H. 1942. A destructive European insect new to the United States. *J. Econ. Ent.* **35:** 289.

86. HALLOCK, H. C. 1936. Life history and control of the Asiatic garden beetle. *Circ. U.S. Dep. Agric.* **246:** 1–20.

87. HIGH, M. M. 1939. The vegetable weevil. *Circ. U.S. Dep. Agric.* **530:** 1–25.

88. HODSON, W. E. H. 1948. Colorado beetle. *N.A.A.S. Quart. Rev.* **1:** 51–2.

89. HOOD, C. E. 1940. Life history and control of the imported willow leaf beetle. *Circ. U.S. Dep. Agric.* **572:** 1–9.

90. HOWARD, L. O. 1930. A history of applied entomology (somewhat anecdotal). *Smithson. Misc. Coll.* **84:** 1–564.

91. HUNTER, W. D. 1926. The pink bollworm . . . *Dep. Bull. U.S. Dep. Agric.* **1397:** 1–30.

92. KRUMHOLZ, L. A. 1948. *Reproduction in the western mosquitofish,* Gambusia affinis affinis (Baird & Girard), *and its use in mosquito control. Ecol. Monogr.* **18:** 1–43.

93. MYERS, G. S. 1940. An American cyprinodont fish, *Jordanella floridae,* reported from Borneo, with notes on the possible widespread introduction of foreign aquarium fishes. *Copeia*: 267–8.

94. NEWELL, W., and BARBER, T. C. 1913. The Argentine ant. *Bull. U.S. Bur. Ent.* **122:** 1–98.

95. PEACE, T. R. 1952. Tree diseases in Great Britain, 1950–51. A general review. *Rep. For. Res., Lond. for . . . 1951*: 94–8.

96. PECHUMAN, L. L. 1938. A preliminary study of the biology of *Scolytus sulcatus* LeC. *J. Econ. Ent.* **31:** 537–43.

97. PHILLIPS, J. C. 1928. Wild birds introduced or transplanted in North America. *Tech. Bull. U.S. Dep. Agric.* **61:** 1–63.

98. ROCKWOOD, L. P. 1926. The clover root borer. *Dep. Bull. U.S. Dep. Agric.* **1426:** 1–48.

99. SCHEFFER, T. H., and COTTAM, C. 1935. The crested myna, or Chinese starling, in the Pacific Northwest. *Tech. Bull. U.S. Dep. Agric.* **467:** 1–26.

100. SMITH, H. S. 1929. On some phases of preventive entomology. *Sci. Mon., N.Y.* **29:** 177–84.

101. SMITH, K. G. 1956. The occurrence and distribution of *Aphomia gularis* (Zell.) (Lep., Galleriidae), a pest of stored products. *Bull. Ent. Res.* **47:** 655–67.

102. SMITH, L. B., and HADLEY, C. H. 1926. The Japanese beetle. *Dep. Circ. U.S. Dep. Agric.* **363:** 1–66.

103. SMITH, M. R. 1936. Distribution of the Argentine ant in the United States

and suggestions for its control and eradication. *Circ. U.S. Dep. Agric.* **387:** 1–39.

104. TROUVELOT, B. 1936. Le dolyphore de la pomme de terre (*Leptinotarsa decemlineata* Say) en Amérique du Nord. *Ann. Épiphyt.* N.S. **1:** 278–336.

105. UNITED STATES BUREAU OF ENTOMOLOGY AND PLANT QUARANTINE. 1941. Japanese beetle. *Insect Pest Surv. Bull. U.S.* **21:** 801–2.

106. VANCE, A. M. 1942. Studies on the prevalence of the European corn borer in the East North Central States. *Circ. U.S. Dep. Agric.* **649:** 1–23.

107. WICHMANN, H. E. 1955. Im europäischen Grossraum eingeschleppte Borkenkäfer. *Z. Angew. Ent.* **37:** 92–109.

108. WICHMANN, H. E. 1957. Einschleppungsgeschichte und Verbreitung des *Xylosandrus germanus* Blandf. in Westdeutschland (nebst einem Anhang: *Xyleborus adumbratus* Blandf.). *Z. Angew. Ent.* **40:** 82–99.

109. WILLIAMS, C. B. 1937. Beware of this beetle! *Zoo,* **11:** 56–7.

110. WILSON, G. FOX. 1935. *Phylloxera* on vines, a new British record. *Proc. R. Ent. Soc. Lond.* **10:** 25–8.

CHAPTER IV

111. ABBOTT, R. T. 1949. March of the giant African snail. *Nat. Hist., N.Y.* **80:** 68–71.

112. AMADON, D. 1950. The Hawaiian honeycreepers (Aves, Drepaniidae). *Bull. Amer. Mus. Nat. Hist.* **95:** 155–262.

113. AURILLIUS, C. 1926. Coleoptera-Curculionidae von Juan Fernandez und der Oster-Insel. In *The natural history of Juan Fernandez and Easter Island* (Ed. C. Skottsberg). Uppsala. **3:** 461–77.

114. BARROW, K. M. 1910. *Three years in Tristan da Cunha.* London.

115. BEQUAERT, J. C. 1950. Studies in the Achatinae, a group of African land snails. *Bull. Mus. Comp. Zool.* **105:** 1–216.

116. BRINCK, P. 1948. Coleoptera of Tristan da Cunha. *Results of the Norwegian Sci. Exped. to Tristan da Cunha 1937–1938,* No. 17: 1–121.

117. BRYAN, E. H. 1940. A summary of the Hawaiian birds. *Proc. 6th Pacif. Sci. Congr., 1939,* **4:** 185–9.

118. BRYAN, W. A. 1915. *Natural history of Hawaii.* Honolulu.

119. BUXTON, P. A., and HOPKINS, G. H. E. 1927. *Researches in Polynesia and Melanesia . . .* Parts I–IV. (*Relating principally to medical entomology.*) *Mem. Lond. Sch. Hyg. Trop. Med.* **1:** 1–260.

120. CHRISTOPHERSEN, E. 1939. Problems of plant geography in Tristan da Cunha. *Norsk Geogr. Tidsskr.* **7:** 106–12. (And personal communication.)

121. CHRISTOPHERSEN, E. 1940. *Tristan da Cunha: the lonely isle.* London.

122. [COOK, J.] [*c.* 1890 ed.] *The three famous voyages of Captain James Cook round the world . . .* London and New York.

123. DAMMERMAN, K. W. 1948. *The fauna of Krakatau 1883–1933.* Amsterdam.

124. DUMBLETON, L. J. 1953. Entomological aspects of insect quarantine in New Zealand. *Proc. 7th Pacif. Sci. Congr., 1949,* **4:** 331–4.

125. ENDERLEIN, G. 1938. Die Dipterenfauna der Juan-Fernandez-Inseln und der Oster-Insel. In *The Natural History of Juan Fernandez and Easter Island* (Ed. C. Skottsberg). Uppsala. **3:** 634–80.

126. GILL, W. W. 1876. *Life in the Southern Isles.* London.

127. GILL, W. W. 1880. *Historical sketches of savage life in Polynesia.* Wellington.

128. GULICK, A. 1932. Biological peculiarities of oceanic islands. *Quart. Rev. Biol.* **7:** 405–27.

129. HAGEN, Y. 1952. Birds of Tristan da Cunha. *Results of the Norwegian Sci. Exped. to Tristan da Cunha 1937–1938*, No. 20: 1–248.

130. JEEKEL, C. A. W. 1954. Diplopoda. *Results of the Norwegian Sci. Exped. to Tristan da Cunha 1937–1938*, No. 32: 5–9.

131. [KING, J. in COOK, J., p. 1010.]

132. MARTIN, J. 1818. *An account of the natives of the Tonga Islands in the South Pacific Ocean . . . compiled from the extensive communications of Mr William Mariner.* London. 2 vols. **1:** Ch. 9.

133. MAYR, E. 1943. The zoogeographic position of the Hawaiian Islands. *Condor,* **45:** 45–8.

134. MUNRO, G. C. 1944. *Birds of Hawaii.* Honolulu.

135. ODHNER, N. H. 1926. Mollusca from Juan Fernandez and Easter Island. In *The natural history of Juan Fernandez and Easter Island* (Ed. C. Skottsberg). Uppsala. **3:** 219–54.

136. PATERSON, C. R. 1953. The establishment and spread in New Zealand of the wasp *Vespa germanica. Proc. 7th Pacif. Sci. Congr., 1949,* **4:** 358–62.

137. PEMBERTON, C. E. 1953. Economic entomology in Hawaii. *Proc. 7th Pacif. Sci. Congr., 1949,* **4:** 91–4.

138. PERKINS, R. C. L. 1903. *Fauna Hawaiiensis or the Zoology of the Sandwich (Hawaiian) Islands.* Vol. I. Part IV. *Vertebrata.* 363–466. Cambridge.

139. PETERSEN, P. ESBEN-. 1924. More Neuroptera from Juan Fernandez and Easter Island. In *The natural history of Juan Fernandez and Easter Island* (Ed. C. Skottsberg). Uppsala. **3:** 309–13.

140. RICHARDS, L. P., and BALDWIN, P. H. 1953. Recent records of some Hawaiian honeycreepers. *Condor,* **55:** 221–2.

141. RIS LAMBERS, D. H. 1955. Aphididae of Tristan da Cunha. *Results of the Norwegian Sci. Exped. to Tristan da Cunha 1937–1938*, No. 34: 1–5.

142. SCHWARTZ, C. W. and E. R. 1949. *A reconnaissance of the game birds in Hawaii.* Hilo, Hawaii.

143. SHERMAN, M., and TAMASHIRO, M. 1956. Biology and control of *Araecerus levipennis* Jordan (Coleoptera: Anthribidae). *Proc. Hawaii. Ent. Soc.* **16:** 138–48.

144. SKOTTSBERG, C. 1920. Notes on a visit to Easter Island. In *The natural history of Juan Fernandez and Easter Island* (Ed. C. Skottsberg). Uppsala. **1:** 1–20.

145. SKOTTSBERG, C. 1927. The vegetation of Easter Island. In *The natural history of Juan Fernandez and Easter Island* (Ed. C. Skottsberg). Uppsala. **2:** 487–502.

146. SKOTTSBERG, C. 1956. Derivation of the flora and fauna of Juan Fernandez and Easter Island. In *The natural history of Juan Fernandez and Easter Island* (Ed. C. Skottsberg). Uppsala. **1**: 193–438.

147. STOKES, J. F. G. 1917. Notes on the Hawaiian rat. *Occ. Pap. Bishop Mus.* **3** (4): 11–21.

148. SVIHLA, A. 1936. The Hawaiian rat. *Murrelet*, **17**: 3–14.

149. TATE, G. H. H. 1935. Rodents of the genera *Rattus* and *Mus* from the Pacific Islands. *Bull. Amer. Mus. Nat. Hist.* **68**: 145–78.

150. VIETTE, P. E. L. 1952. Lepidoptera. *Results of the Norwegian Sci. Exped. to Tristan da Cunha 1937–1938*, No. 23: 1–19.

151. VON HOCHSTETTER, F. 1867. *New Zealand: its physical geography, geology and natural history.* Stuttgart.

152. WEBER, P. W. 1956. Recent introductions for biological control in Hawaii. 1. *Proc. Hawaii. Ent. Soc.* **16**: 162–4.

153. WHEELER, W. M. 1934. Revised list of Hawaiian ants. *Occ. Pap. Bishop Mus.* **10** (21): 1–21.

154. WILLIAMS, F. X. 1953. Some natural enemies of snails of the genus *Achatina* in East Africa. *Proc. 7th Pacif. Sci. Congr., 1949*, **4**: 277–8.

155. WILSON, S. B., and EVANS, A. H. 1890–99. *Aves Hawaiiensis: the birds of the Sandwich Islands.* London.

156. WODZICKI, K. A. 1950. Introduced mammals of New Zealand: an ecological and economic survey. *Bull. D.S.I.R., N.Z.* **98**: 1–255.

157. ZIMMERMANN, A. 1924. Coleoptera-Dytiscidae von Juan Fernandez und der Oster-Insel. In *The natural history of Juan Fernandez and Easter Island* (Ed. C. Skottsberg). Uppsala. **3**: 298–304.

158. ZIMMERMAN, E. C. 1948. *Insects of Hawaii.* **1.** *Introduction.* Honolulu.

CHAPTER V

159. ALLEN, K. R. 1956. The geography of New Zealand's freshwater fish. *N.Z. Sci. Rev.* **14** (3): 3–9.

160. ANON. 1940. [Introduction of *Nereis* into the Caspian Sea.] *The Times*, 7 November.

161. BISHOP, M. W. H. 1951. Distribution of barnacles by ships. *Nature*, **167**: 531.

162. BISHOP, M. W. H., and CRISP, D. J. 1957. The Australasian barnacle, *Elminius modestus*, in France. *Nature*, **179**: 482–3.

163. CALHOUN, A. J. 1952. Annual migrations of California striped bass. *Calif. Fish Game*, **38**: 391–403.

164. CALMAN, W. T. 1927. Zoological results of the Cambridge Expedition to the Suez Canal, 1924. **XIII.** Report on the Crustacea Decapoda (Brachyura). (With Appendix by H. MUNRO FOX.) *Trans. Zool. Soc. Lond.* **22**: 211–19.

165. COLE, H. A. 1942. The American whelk tingle, *Urosalpinx cinerea* (Say), on British oyster beds. *J. Mar. Biol. Ass. U.K.* **25**: 477–508.

166. COLE, H. A. 1952. The American slipper limpet (*Crepidula fornicata* L.) on Cornish oyster beds. *Fish. Invest., Lond.* Ser. 2, **17**(7): 1–13.

167. COLE, H. A. 1956A. Benthos and the shellfish of commerce. In *Sea fisheries: their investigation in the United Kingdom* (Ed. M. Graham). London. 139–206.

168. COLE, H. A. 1956B. *Oyster cultivation in Britain. A manual of current practice.* London.

169. COLE, H. A., and BAIRD, R. H. 1953. The American slipper limpet (*Crepidula fornicata*) in Milford Haven. *Nature*, **172**: 687.

170. CONNELL, J. H. 1955. *Elminius modestus* Darwin, a northward extension of range. *Nature*, **175**: 954.

171. CRISP, D. J., and CHIPPERFIELD, P. N. J. 1948. Occurrence of *Elminius modestus* (Darwin) in British waters. *Nature*, **161**: 64.

172. DAVIDSON, F. A., and HUTCHINSON, S. J. 1938. The geographic distribution and environmental limitations of the Pacific salmon (genus *Onchorhynchus*). *Bull. U.S. Bur. Fish.* **48**: (No. 26) 667–92.

173. EDMONDSON, C. H., and WILSON, I. H. 1940. The shellfish resources of Hawaii. *Proc. 6th Pacif. Sci. Congr., 1939*, **3**: 241–3.

174. EKMAN, S. 1953. *Zoogeography of the sea.* London.

175. FELDMAN, J. and G. 1942. Recherches sur les Bonnemaisoniacées et leur alternance de genérations. *Ann. Sci. Nat. Bot.* Ser. 11, **3**: 75–175.

176. FOX, H. MUNRO. 1926. Zoological results of the Cambridge Expedition to the Suez Canal, 1924. I. General part. *Trans. Zool. Soc. Lond.* **22**: 1–64.

177. HARDY, A. C. 1956. *The open sea. Its natural history: the world of plankton.* London.

178. HENTSCHEL, E. 1923. Der Bewuchs an Seeschiffen. *Int. Rev. Hydrobiol.* **11**: 238–64.

179. HILDEBRAND, S. F. 1939. The Panama Canal as a passageway for fishes, with lists and remarks on the fishes and invertebrates observed. *Zoologica, N.Y.* **24**: 15–45.

180. HORRIDGE, G. A. 1951. Occurrence of *Asparagopsis armata* Harv. on the Scilly Isles. *Nature*, **167**: 732–3.

181. JOHNSON, W. C., and CALHOUN, A. J. 1952. Food habits of California striped bass. *Calif. Fish Game*, **38**: 531–4.

182. JONES, L. L. 1940. An introduction of an Atlantic crab into San Francisco Bay. *Proc. 6th Pacif. Sci. Congr., 1939*, **3**: 485–6.

183. KINCAID, T. 1953. The acclimatization of the Pacific oyster (*Ostrea laperousii* Schrenck = *Ostrea gigas* Thunberg) upon the west coast of North America. *Proc. 7th Pacif. Sci. Congr., 1949*, **4**: 508–12.

184. LOOSANOFF, V. L. 1955. The European oyster in American waters. *Science*, **121**: 119–21.

185. MANSUETI, R., and KOLB, H. 1953. A historical review of the shad fisheries of North America. *Publ. Chesapeake Biol. Lab.* **97**: 1–293.

186. MARSHALL, N. B. 1952. The 'Manihine' Expedition to the Gulf of Aqaba 1948–1949. IX. Fishes. *Bull. Brit. Mus. (Nat. Hist.) Zool. Ser.* **1**: 221–52.

187. MERRIMAN, D. 1941. Studies on the striped bass (*Roccus saxatilis*) of the Atlantic coast. *Fish. Bull., U.S. Fish & Wildlife Service*, **50** (No. 35): 1–77.

188. MISTAKIDIS, M. N., and HANCOCK, D. A. 1955. Reappearance of *Ocenebra erinacea* (L.) off the east coast of England. *Nature*, **175**: 734.

189. NEAVE, F. 1954. Introduction of anadromous fishes on the Pacific coast. *Canad. Fish Culturist*, No. 16: 25–6.

190. NIKITIN, V. N. (Ed. by). 1952. [Miscellany on the introduction of *Nereis succinea* into the Caspian Sea.] *Material on Fauna and Flora, N.S., Zool. Sect.*, No. 33 (48): 1–372. (In Russian.)

191. ORTON, J. H. 1937. *Oyster biology and oyster-culture, being the Buckland Lectures for 1935*. London.

192. REES, C. B., and CATTLEY, J. G. 1949. *Processa aequimana* Paulson in the North Sea. *Nature*, **164**: 367.

193. SANDISON, E. E. 1950. Appearance of *Elminius modestus* in South Africa. *Nature*, **165**: 79–80.

194. SCOFIELD, N. B., and BRYANT, H. C. 1926. The striped bass in California. *Calif. Fish Game*, **12**: 55–74.

195. SLOCUM, J. 1948. *Sailing alone around the world*. London. (Originally, publ. 1900.)

196. SMITH, H. M. 1895. A review of the history and results of the attempts to acclimatize fish and other water animals in the Pacific states. *Bull. U.S. Fish. Comm.* **15**: 379–472.

197. STEINITZ, W. 1929. Die Wanderung indopazifischer Arten ins Mittelmeer seit Beginn der Quartärperiode. *Int. Rev. Hydrobiol.* **22**: 1–90.

198. THOMPSON, J. M. 1952. The acclimatization and growth of the Pacific oyster (*Gryphaea gigas*) in Australia. *Aust. J. Mar. Freshw. Res.* **3**: 64–73.

199. WALKER, M. I., BURROWS, E. M., and LODGE, S. M. 1954. Occurrence of *Falkenbergia rufolanosa* in the Isle of Man. *Nature*, **174**: 315.

200. WALNE, P. R. 1956. The biology and distribution of the slipper limpet *Crepidula fornicata* in Essex rivers with notes on the distribution of the larger epi-benthic invertebrates. *Fish. Invest., Lond.* Ser. 2, **20** (6): 1–50.

201. WOLFF, T. 1954A. Tre østamerikanske krabber fundet i Danmark. *Flora og Fauna*, **60**: 19–34.

202. WOLFF, T. 1954B. Occurrence of two East American species of crabs in European waters. *Nature*, **174**: 188–9.

203. ZENKEVICH, L. A. 1937. [Progress in the study of the marine fauna of the U.S.S.R. made in twenty years.] *Zool. Zh.* **16**: 830–70. (In Russian.)

CHAPTER VI

204. BROWN, R. C., and SHEALS, R. A. 1944. The present outlook on the gypsy moth problem. *J. For.* **42**: 393–407.

205. CAIN, A. J., and CUSHING, D. H. 1948. Second occurrence and persistence of the amphipod *Orchestia bottae* M. Edwards in Britain. *Nature*, **161**: 483.

206. COLE, H. A. 1952. The American slipper limpet (*Crepidula fornicata* L.) on Cornish oyster beds. *Fish. Invest., Lond.* Ser. 2, **17** (7): 1–13.

207. CRAWFORD, G. I. 1937. A review of the amphipod genus *Corophium*, with notes on the British species. *J. Mar. Biol. Ass. U.K.* **21:** 589–630.

208. DRUCE, G. C. 1897. *The flora of Berkshire.* Oxford.

209. ELTON, C. S., and MILLER, R. S. 1954. The ecological survey of animal communities: with a practical system of classifying habitats by structural characters. *J. Ecol.* **42:** 460–96.

210. FORBUSH, E. H., and FERNALD, C. H. 1896. *The gypsy moth.* Porthetria dispar (*Linn.*). Boston.

211. FRYER, G. 1951. Distribution of British freshwater Amphipoda. *Nature,* **168:** 435.

212. GODWIN, H. 1956. *The history of the British flora: a factual basis for phytogeography.* Cambridge.

213. HULL, T. G. 1941. *Diseases transmitted from animals to man.* Springfield and Baltimore.

214. HYNES, H. B. N. 1956. British freshwater shrimps. *New Biology,* **21:** 25–42. Harmondsworth, Middlesex.

215. KEANE, C. 1926. The epizootic of foot and mouth disease. *Spec. Publ. Calif. Dep. Agric.* **65:** 1–54.

216. KING, W. B. R., and OAKLEY, K. P. 1936. The Pleistocene succession in the lower parts of the Thames Valley. *Proc. Prehist. Soc.,* N.S. **2:** 52–76.

217. MCCUBBIN, W. A. 1954. The plant quarantine problem. *Ann. Cryptog. Phytopath., Copenhagen,* **11:** 1–255.

218. MAXWELL, H. 1915. Waterfowl and the American pond-weed (*Elodea canadensis*). *Scot. Nat.*: 81–3.

219. MISTAKIDIS, M. N. 1951. Quantitative studies of the bottom fauna of Essex oyster grounds. *Fish. Invest., Lond.* Ser. 2, **17:** (6): 1–47.

220. MUNRO, T. 1935. Note on musk-rats and other animals killed since the inception of the campaign against musk-rats in October 1932. *Scot. Nat.*: 11–16.

221. PRATT, A. (Nineteenth century.) *The flowering plants and ferns of Great Britain.* London. **5.**

222. REID, D. M. 1948. Occurrence of the amphipod *Orchestia bottae* and other organisms in Britain. *Nature,* **161:** 609.

223. RIDLEY, H. N. 1930. *The dispersal of plants throughout the world.* Ashford, Kent.

224. SHERMAN, R. W. 1956. New trends in plant quarantine. *J. Econ. Ent.* **49:** 881–3.

225. SHORTEN, M. 1954. *Squirrels.* London.

226. SMITH, H. S. *et al.* 1933. The efficiency and economic effects of plant quarantines in California. *Bull. Calif. Agric. Exp. Sta.* **553:** 1–276.

227. SPARKS, B. W. 1957. The non-marine Mollusca of the Interglacial deposits at Bobbitshole, Ipswich. *Phil. Trans.* Ser. B, **241:** 1–44.

228. SPOONER, G. M. 1951. Distribution of British freshwater Amphipoda. *Nature,* **167:** 530.

229. UDINE, E. J. 1941. The black grain stem sawfly and the European wheat stem sawfly in the United States. *Circ. U.S. Dep. Agric.* **607**: 1–9.
230. WARWICK, T. 1934. The distribution of the muskrat (*Fiber zibethicus*) in the British Isles. *J. Anim. Ecol.* **3**: 250–67.
231. WARWICK, T. 1941. A contribution to the ecology of the musk-rat (*Ondatra zibethica*) in the British Isles. *Proc. Zool. Soc. Lond.* Ser. A, **110**: 165–201.

CHAPTER VII

232. ABBOTT, R. T. 1951. Operation snailfolk. *Nat. Hist., N.Y.* **60**: 280–5.
233. ALLEN, A. A. 1953. *Graphocephala coccinea* Forst. (Hem., Tettigoniellidae) in Kent, etc. *Ent. Mon. Mag.* **89**: 71.
234. BABERS, F. H. 1949. Development of insect resistance to insecticides. *U.S. Bur. Ent. Plant Quarant.* E-776: 1–31.
235. BABERS, F. H., and PRATT, J. J. 1951. Development of insect resistance to insecticides. II. A critical review of the literature up to 1951. *U.S. Bur. Ent. Plant Quarant.* E-818: 1–45.
236. BAILLIE, A. F. H., and JEPSON, W. F. 1951. Bud blast disease of the *Rhododendron. J. R. Hort. Soc.* **76**: 355–65.
237. BROWN, J. M. B. 1954. *Rhododendron ponticum* in British woodlands. *Rep. For. Res., Lond.* . . . 1953: 42–3.
238. CLAUSEN, C. P., KING, J. L., and TERANISHI, C. 1927. The parasites of *Popillia japonica* in Japan and Chosen (Korea) and their introduction into the United States. *Dep. Bull. U.S. Dep. Agric.* **1429**: 1–55.
239. CULLINAN, F. P. 1949. Some new insecticides—their effect on plants and soils. *J. Econ. Ent.* **42**: 387–91.
240. DAVIDSON, G. 1956. Insecticide resistance in *Anopheles gambiae* Giles: a case of simple Mendelian inheritance. *Nature,* **178**: 863–4.
241. DEBACH, P., FLESCHNER, C. A., and DIETRICK, E. J. 1951. A biological check method for evaluating the effectiveness of entomophagous insects. *J. Econ. Ent.* **44**: 763–6.
242. DODD, A. P. 1940. *The biological campaign against prickly-pear.* Commonwealth Prickly Pear Board. Brisbane.
243. DODGE, B. O., and RICKETT, H. W. 1943. *Diseases and pests of ornamental plants.* Lancaster, Pennsylvania.
244. DUFFY, E. A. J. 1953. *A monograph of the immature stages of British and imported timber beetles.* London.
245. ELTON, C. 1927. *Animal ecology.* London.
246. FOSTER, A. C. 1951. Some plant responses to certain insecticides in the soil. *Circ. U.S. Dep. Agric.* **862**: 1–41.
247. FRIEDMANN, H. 1929. *The cowbirds. A study in the biology of social parastism.* Springfield and Baltimore.
248. GARRETSON, M. S. 1938. *The American bison.* New York.
249. GLÜCK, G. 1939. *Pieter Brueghel le Vieux.* Paris.

250. GORIUS, U. 1955. Untersuchungen über den Lärchenbock, *Tetropium Gabrieli* Weise mit besonderer Berücksichtigung seines Massenwechsels. *Z. Angew. Ent.* **38:** 157–205.

251. HANNA, A. D., JUDENKO, E., and HEATHERINGTON, W. 1956. The control of *Crematogaster* ants as a means of controlling the mealybugs transmitting the swollen-shoot virus disease of cacaó in the Gold Coast. *Bull. Ent. Res.* **47:** 219–26.

252. HEWITT, C. G. 1921. *The conservation of the wild life of Canada.* New York.

253. HOWES, F. N. 1945. *Plants and beekeeping.* London.

253*a*. JONES, C. GARRETT-. 1954. Evidence of the development of resistance to DDT by *Anopheles sacharovi* in the Levant. *Bull. World Hlth Org.* **11:** 865–83.

254. KLEMENT, Z., and KIRÁLY, Z. 1957. Hyperparasitic chain of a fungus, a bacterium and its phage on wheat. *Nature,* **179:** 157–8.

255. LEOPOLD, A. 1949. *A sand county almanac, and sketches here and there.* New York.

256. LEOPOLD, A. S., and DARLING, F. F. 1953. *Wildlife in Alaska: an ecological reconnaissance.* New York.

257. LOTKA, A. J. 1925. *Elements of physical biology.* Baltimore.

258. LOTKA, A. J. 1927. Fluctuations in the abundance of species considered mathematically. [With comment by V. VOLTERRA.] *Nature,* **119:** 12–13.

259. METCALFE, H., and COLLINS, J. F. 1911. The control of the chestnut bark disease. *Farmers' Bull. U.S. Dep. Agric.* **467:** 1–24.

260. PALMER, L. J., and ROUSE, C. H. 1945. Study of the Alaska tundra with reference to its reactions to reindeer and other grazing. *Res. Rep. U.S. Fish & Wildlife Service,* **10:** 1–48.

261. ROE, F. G. 1951. *The North American buffalo: a critical study of the species in its wild state.* Toronto.

262. SANSOME, F. W. 1940. Breeding disease-resistant plants. *Nature,* **145:** 690–3.

263. SHEALS, J. G. 1956. Soil population studies. I.—The effects of cultivation and treatment with insecticides. *Bull. Ent. Res.* **47:** 803–22.

264. SMITH, H. S. *et al.* 1933. The efficacy and economic effects of plant quarantines in California. *Bull. Calif. Agric. Exp. Sta.* **553:** 1–276.

265. THOMAS, F. J. D. 1957. The residual effects of crop-protection chemicals in the soil. In *Plant Protection Conference 1956. Proceedings of the Second International Conference at Fenhurst Research Station, England.* London. 215–22.

266. TILDEN, J. W. 1951. The insect associates of *Baccharis pilularis* De Candolle. *Microentomology,* **16:** 149–88.

267. VOGT, W. 1948. *Road to survival.* New York.

268. VOLTERRA, V. 1931. *Leçons sur la théorie mathématique de la lutte pour la vie.* Paris.

269. WARBURG, E. F. 1953. A changing flora as shown in the status of our trees and shrubs. In *The changing flora of Britain* (Ed. J. E. LOUSLEY). Arbroath. 171–80.

270. WATSON, J. S. 1951. The rat problem in Cyprus. *Colonial Res. Publ.* **9:** 1–66.

271. WILSON, G. FOX. 1939A. Insect pests of the genus *Rhododendron*. *Proc. 7th Int. Congr. Ent.*, *1938*, **4:** 2296–2323.
272. WILSON, G. FOX. 1939B. Insect pests of rhododendrons: their distribution in Britain. *Proc. R. Ent. Soc. Lond.* Ser. A, **14:** 1–5.
273. WILSON, G. FOX. 1950. Pests of flowers and shrubs. *Bull. Minist. Agric., Lond.* **97:** 1–105.
274. WORTHINGTON, S. and E. B. 1933. *Inland waters of Africa.* London.

CHAPTER VIII

275. AUDY, J. R. 1956. Ecological aspects of introduced pests and diseases. *Med. J. Malaya,* **11:** 21–32.
276. COLLYER, E. 1953A. Biology of some predatory insects and mites associated with the fruit tree red spider mite (*Metatetranychus ulmi* (Koch)) in south-eastern England. II. Some important predators of the mite. *J. Hort. Sci.* **28:** 85–97.
277. COLLYER, E. 1953B. Ibid. III. Further predators of the mite. *J. Hort. Sci.* **28:** 98–113.
278. DEBACH, P., FLESCHNER, C. A., and DIETRICK, E. J. 1949. Natural control of the California red scale in untreated orchards in southern California. *Proc. 7th Pacif. Sci. Congr.*, *1949*, **4:** 236–48.
279. GAUSE, G. F. 1934. *The struggle for existence.* Baltimore.
280. GAUSE, G. F. 1935. Vérification expérimentales de la théorie mathématique de la lutte pour la vie. *Actualités Sci. Industr.* No. 277: 1–63.
281. GAUSE, G. F., SMARAGDOVA, N. P., and WITT, A. A. 1936. Further studies of interaction between predators and prey. *J. Anim. Ecol.* **5:** 1–18.
282. LORD, F. T. 1947. The influence of spray programmes on the fauna of apple orchards in Nova Scotia: II. Oystershell scale *Lepidosaphes ulmi* (L.). *Canad. Ent.* **79:** 196–209.
283. LORD, F. T. 1956. Ibid. IX. Studies on means of altering predator populations. *Canad. Ent.* **88:** 129–37.
284. PICKETT, A. D. 1949. A critique on insect chemical control methods. *Canad. Ent.* **81:** 67–76.
285. PICKETT, A. D. 1953. Controlling orchard insects. *Agric. Inst. Rev.* **8:** 52–3.
286. PICKETT, A. D., and PATTERSON, N. A. 1953. The influence of spray programmes on the fauna of apple orchards in Nova Scotia. IV. Review. *Canad. Ent.* **85:** 472–8.
287. SCHWEITZER, A. 1956. *My life and thought: an autobiography.* (Transl. C. T. Campion.) London. 188.
288. SMITH, H. S., and DEBACH, P. 1953. Artificial infestation of plants with pest insects as an aid in biological control. *Proc. 7th Pacif. Sci. Congr., 1949,* **4:** 255–9.
289. VOÛTE, A. D. 1946. Regulation of the density of the insect-populations in virgin-forests and cultivated woods. *Arch. Néerl. Zool.* **7:** 435–70.

CHAPTER IX

290. ČAPEK, K. 1925. *Letters from England.* London.
291. CHANT, D. A. 1956. Predacious spiders in orchards in south-eastern England. *J. Hort. Sci.* **31**: 35–46.
292. FORESTRY COMMISSION. 1955. *Report of the Committee on Hedgerow and Farm Timber 1955.* London.
293. GRAHAM, E. H. 1944. *Natural principles of land use.* New York, etc.
293*a*. GRAHAM, E. H. 1957. Nature protection as a part of land development. *Proc. 6th Meeting Int. Union Conserv. Nature & Nat. Res.*: 194–201.
294. HARTKE, W. 1951. Die Heckenlandschaft. Der geographische Charakter eines Landeskulturproblems. *Erdkunde*, **5**: 132–52.
295. JEFFRIES, R. 1879. *Wild life in a southern county.* London. Ch. 3.
296. MARQUARDT, G. 1950. Die Schleswig-Holsteinische Knicklandschaft. *Schr. Geogr. Inst. Univ. Kiel*, **13** (3): 1–90.
297. MUIR, J. 1909. *Our National Parks.* Boston and New York.

Index